Mysteries of the
Solar System

R. A. Lyttleton

Fellow of St. John's College, Cambridge

Mysteries of the Solar System

CLARENDON PRESS · OXFORD
1968

Oxford University Press, Ely House, London W.1

GLASGOW NEW YORK TORONTO MELBOURNE WELLINGTON
CAPE TOWN SALISBURY IBADAN NAIROBI LUSAKA ADDIS ABABA
BOMBAY CALCUTTA MADRAS KARACHI LAHORE DACCA
KUALA LUMPUR HONG KONG TOKYO

MADE AND PRINTED IN GREAT BRITAIN
BY WILLIAM CLOWES AND SONS LTD
LONDON AND BECCLES

Preface

THESE essays are based upon a series of lectures given at Brandeis University, Waltham, Massachusetts, during my visit as Jacob Siskind Professor of Astronomy in the autumn of 1965. It was the wish of Brandeis University that the lectures should be published as some record of my tenure of that Chair, and the present volume is the result.

The essays published here are more detailed and more extensive than the original lectures, which were limited by the time available. It should nevertheless be emphasized that even these expanded discussions are not claimed to be in any way exhaustive: the topics considered can scarcely be more than broached in the space of an hour's talk or in a single chapter of a book. The essays may, however, serve as an introduction to the problems involved. In the main, they present views and attitudes to these problems that have gradually come to seem to me to be appropriate, though some of them run counter to conclusions that are, I have found, fairly widely regarded as well established.

In entitling the lectures and the book 'Mysteries of the Solar System', my aim was to remind both audience and readers how little is known for certain about even the nearest realm of the universe, namely our own solar system. The advent of space-research has already resulted in many remarkable achievements and new astronomical discoveries about this complex system, though only the merest beginning has as yet been made with the development of techniques that must surely have limitless possibilities and permanent advantages over methods restricted to the locality of the Earth itself. Even so, progress is likely to be still more rapid the more closely experiment and observation are related to and guided by theory. The more clearly we recognize the extent and limitations of our knowledge, the more readily can we decide which areas require further study, and the more easily can we perceive valuable lines of enquiry.

In other words, the theoretical status of every observation and conclusion should be properly appreciated at every stage of research, and constantly re-assessed in the light of new observations and theoretical advances. One purpose of these essays is to make a small contribution to this essential aspect of the scientific method in relation to some problems of the solar system that I happen to have studied.

St. John's College
Cambridge
June 1967

R. A. L.

Contents

List of Plates

the great central problems of astronomy. It is one with which little solid progress has been made, despite the fact that much has been written on it. The problem stands as a kind of permanent challenge, for we cannot even be sure whether, at any particular stage of our understanding of astronomy, we are in a position to be able to solve it. Nevertheless try we must, for successive attempts on the problem have brought to light areas of thought that it has become evident would need to be far better understood before the actual problem could be tackled with any hope of success.

For example, it is obvious today that any attempt 400 years ago to determine the origin of the solar system would have been doomed to failure because there would have been no knowledge of Newtonian dynamics. Without this all-important theory, it would not have been possible to assess and formulate what precisely the problem might be, and decide the relative importance of various factors. For a long period of history, brilliant comets with tails stretching across the sky were generally regarded as the most important of all celestial objects, and until gravitational theory became available there was no way of demonstrating their relative insignificance astronomically.

Where the specific problem of the origin of the solar system is concerned, and indeed for any scientific problem, there is the prior question of what would constitute a satisfactory theory. Experience shows that this is to a large extent a subjective matter: a theory that satisfies one person may fail altogether to satisfy another. In the last analysis it is an aesthetic matter to what extent a solution of a scientific problem may be regarded as valid. To many people in past ages, the so-called apologues of Genesis have seemed to constitute a perfectly acceptable theory of the universe. Even to this day some people wonder why scientists cannot be satisfied with such explanations, why they cannot rest content with this account of how everything came about and happened. Why should it not be the case that the whole universe was created in the year 4004 B.C. by some omnipotent power, or indeed created just a few moments ago?

It is true, of course, that it would be necessary to create it with rocks looking as if they were thousands of millions of years old, and with a good many other signs of a remote antiquity, and ourselves with all our memories of the past stored in our minds. But why not? If one postulates *omnipotence*, then it is omnipotent, and there is no reason why all these things should not be accomplished thereby.

Now the reason a scientist feels dissatisfaction with this sort of approach to the problem is that it produces no insight into the world, it gives no idea of how the world can have come about, and above all it has no predictive power. For example, a seemingly simple question often asked is whether there are any other planets in the universe circling round other suns. This is not a question that one could answer if one accepted the idea of an instantaneous creation. On the other hand, if a scientific theory showing the processes by which planets come into existence were available, it would automatically give information on the likelihood of planets having formed elsewhere.

At present there is renewed interest in the question of whether intelligent life on planets similar to the Earth exists anywhere else in the universe, but there is no way of deciding even whether suitable planetary abodes exist until we have a theory of the origin of the planets. Any possibility of inspecting other stars telescopically to see if they have planets is easily seen to be quite hopeless at the present time. If we imagine the great planet Jupiter, which is intrinsically the brightest of the planets, removed to the distance of the nearest star α-Centauri, which is a body fairly similar to the sun, instead of appearing to us as an object of luminosity magnitude − 2, it would drop down to something less than magnitude 22, which is about the limit that the 200-inch telescope can detect with long exposure. So the chance of seeing even a 'Jupiter' at that distance is remote indeed, and a tiny 'Earth' less still. But the actual situation would be a great deal worse even than this, because within 2″ or 3″ (seconds of arc) of the sought-for planet there would be a brilliant star thousands of millions of times brighter, and this would flood the field with its own light. So there would be no possibility of detecting such a planet optically even if associated with the *nearest* star. For remoter stars the situation gets progressively more hopeless, particularly since what people really want to know when they ask this question is not whether other stars have other mighty Jupiters going round them but whether they have tiny Earths similar to our own and capable of supporting life.

But if we had a reliable theory of the origin of planets, if we knew of some mechanism consistent with the laws of physics so that we *understood* how planets form, then clearly we could make use of it to estimate the probability that other stars have attendant planets. However, no such theory exists yet, despite the large number of hypotheses suggested. At one time a much-discussed hypothesis was that of another star actually colliding with the sun, an event so extremely unlikely, because of the vast distances that separate stars compared with their bodily dimensions, that it would have implied that our sun might even be unique in the galaxy in possessing planets. This hypothesis has long since been shown to fail for dynamical reasons, but it serves to illustrate how once one has a hypothesis one is in a position to assess the possibility of there being other planetary systems formed in accordance with processes based upon it.

The nature of the problem of the origin of the solar system is a curious one from the point of view of scientific method because it is a theory of the past, and although we tend to regard past events as certain, especially things we believe we have witnessed ourselves, yet they are completely unverifiable. There is no way that we can go back and see what happened, and we can only hope to make theoretical constructs relating to our views about the past but capable of being tested at the present time. For a theory of the solar system to be satisfactory, it would have to account for all existing known properties of the system. If in addition it predicted other unexpected properties found to be true, the theory would be of great heuristic value. Then again,

one of the desires that urge people to struggle with this problem
is that they wish to feel sure that the solar system can happen
within the laws of physics. They wish to feel sure that no 'special',
super-natural, event has to be postulated; to feel that everything is 'all right', and that the laws of science are sufficiently
comprehensive to allow the solar system to happen. In other
words, the search for a satisfactory theory constitutes a check
on dynamics and physics, and scientists will only be able to feel
comfortable in their minds about the solar system when this
has been demonstrated in full. Until the mystery is entirely
cleared up, scientists cannot really quite relax.

In order to tackle the problem itself, we have first to try to
decide what features are really significant, to decide what can
be relied upon as trustworthy clues to the origin. It is not
necessary to recount again how, from the phenomenon of
radioactivity, the great but nevertheless finite age of the system
has been established. We are now fairly certain that the planets
have existed for something like 4 to 5 thousand million years,
four to five aeons (to use a modern unit of time, the *aeon*, which
avoids the confusion associated with the word *billion*). This,
then, is the order of the time-scale of the solar system, and we
are examining it and studying it only after there has been this
vast interval to alter and efface original features.

So the first problem is to try to decide what the original
features may have been, for there is no point in building an
elaborate theory of the formation of planets to explain some
apparently clear-cut present-day property of the system if in
fact that property is not really original at all but has come
about otherwise. For example, a great many theories have
concentrated for their starting-point on the fact that the
planetary orbits are very nearly circular and almost confined
to the same plane. But our knowledge of the consequences of
dynamical forces over periods of time measured in aeons is nil;
and moreover we do not yet know what subsequent dissipative
actions may have been operative since the origin of the system;
so we simply do not know whether this very obvious property

was original to it or not, and we are unlikely to know until we have a satisfactory theory. Again, it has been noted by several theorists that there is an empirical law of progression in the distances of successive planets from the sun, a 'law' associated with the name of Bode. Roughly, it is that going outwards from the sun to the planets in order—Mercury, Venus, the Earth, Mars, Jupiter, Saturn, Uranus, and Neptune—each one is at about twice the distance of the previous one. Remarkable as the law may be, it is very difficult to assess its importance and status. My own view is that too much significance may have been attached to Bode's law. In fact it fails rather badly in the inner reaches of the solar system, and on the one occasion when it might really have helped—the discovery of Neptune (which forms the subject of Chapter 7)—it placed the planet at nearly 40 AU from the sun, forty times the Earth's distance, when the true distance is only about 30 AU. But it is essential that the status of such a law should be understood before it is worth while developing a theory that assumes this distribution to be the original arrangement of the planets continuing unchanged to the present day.

When one considers the material content of the solar system, which presumably at least in order of magnitude must be original to it, the list is long and impressive. Besides the nine major planets (if we include Pluto as a planet, though it may be an escaped satellite of Neptune), there are thirty-one satellites at present known associated gravitationally with the planets. It is estimated that there are more than 30 000 asteroids, in planetary motion round the sun, capable of being detected with existing telescopes. The known asteroids lie mainly between the orbits of the Earth and Jupiter, but only about a tenth of these have actually been found, and it is not even possible to guess whether other asteroidal belts exist farther out than Jupiter. But there is little doubt from other considerations that even within Jupiter's distance there must be countless millions of tinier asteroids still, in the form of meteorites, that could never be detected optically. The Earth itself is bombarded by a few of these every day, but their frequency in planetary space is quite uncertain as yet. But returning to the list of contents, there are something of the order of a million comets gravitationally bound to the sun, and there are meteor-streams containing myriads of tiny particles associated with some of these comets, all in orbital motion round the sun. And there is interplanetary dust and interplanetary gas.

Although over a few hundred or thousand years the sum total of these can have very little effect on the motions of the planets, when it comes to aeons of time we can at present form little idea of what their systematic effects may have been. There simply is not available adequate dynamical and physical theory to investigate such questions even if the actions have been continuous and gradual. But in addition there are types of effects for which we could never make allowance at all. Thus it is known from inspection of the surface of the moon that it has been subjected to a great many violent meteoritic impacts, and even if we were able to run the theory of the motion of the moon backwards in time for thousands of millions of years (which we cannot) in order to allow for these impacts, it would also be necessary to know the precise circumstances of their occurrence, and to reverse these in order to follow the motion backwards. Not only would this be impossible because of the enormous number of such events, but the necessary details are obviously unknowable.

The effect of interstellar dust and gas over long periods of time may also have been considerable, but here again little is known of the present distribution or of how this may have varied in the past. As the sun pursues its orbit in the galaxy, it must from time to time pass through interstellar gas and dust clouds. Something like 10 per cent of the volume of the galaxy is occupied by such clouds, and accordingly for about 10 per cent of its lifetime the sun and planets will be moving through such material. There must necessarily be intense gravitational reaction. The nature of this reaction is difficult to work out theoretically, but it may well be that from time to time the sun

2—M.S.S.

captures an enormously extended gaseous atmosphere reaching out to planetary distances, and this might well have highly important dissipative effects on both the planetary and satellite motions.

It is obvious that within the solar system there is a traffic problem despite the fact that there is plenty of room, and there is not the slightest doubt that meteoritic collisions have occurred with other bodies besides the moon. In particular, the Earth itself must have been at least as heavily bombarded, area for area, as the moon. There is growing evidence for the existence of numerous terrestrial craters, but the great rapidity of weathering processes renders discovery of the evidence difficult. When the earlier dynamical astronomers, led by Newton, were first beginning the work of surveying the solar system, its great age was not in the least appreciated. Through obscurant influences on knowledge, it had come to be thought that the whole previous activity of the universe could be compressed into a mere few thousand years, and the hope of these early dynamical astronomers, at long last in possession of the laws of motion, was that they could simply insert the present positions and velocities of the planets, and work backwards in time to reveal the initial state of the system. We can see now how hopelessly doomed to failure this attractive plan was, but the thrilling prospect urged them on to lay the foundations and pattern for the whole of physical science since. All this work was, of course, carried out on the tacit assumption that the then known planets constituted the entire system—dynamical theories were sometimes published under the impressive title of 'System of the World'—and there was not the slightest suspicion of the existence of the many other forms of material in the solar system besides the observed planets and comets. But we can see clearly today how little hope there is in deciding what the earliest dynamical state of the system may have been, and almost equally so we cannot decide its future state for lengths of time measured in aeons.

Because of the extreme awkwardness of the mathematical

equations involved, the dynamical theory of the motion of the planets proceeds by successive approximations, and takes advantage of the fact that the masses of the planets are all small compared with the mass of the central main attracting body, the sun, which may be taken as having mass unity. Even for the largest planet, Jupiter, the mass is only about one-thousandth, and the fact that this is a small quantity is important mathematically. For most of the others, the mass is far less: for example, our Earth is only about three-millionths the mass of the sun. The working out of the perturbations that the planets produce upon each other is extremely arduous, and it has been carried out as a literal procedure, that is in algebraic and trigonometrical form, only to the second order in the masses, which means as far as the squares of these small quantities. Each succeeding approximation may involve far more labour than the preceding ones put together, and computing techniques have yet to be developed that would extend even to the third order in the masses. But as far as the theory has yet been taken, there is no sign of any fundamental change occurring in the solar system. The major axes of the orbits vary only by small purely periodic terms, but none that would imply any general change in the mean distances of the planets from the sun. Then again the eccentricities merely oscillate slightly between fixed limits, and the upper limits reached are small. The orbits as a whole can slew round in space: systematic motions of the perihelion points and of the nodes occur: but these in no way imply any instability or overall change in the general arrangement of the system. On the other hand, because of the limitations imposed by the degree of approximation, the theory can apply only for a limited period of time, and there is no guarantee whatever that the theory can be trusted for much more than a million or possibly ten million years.

Thus the most that we know from dynamical theory, taking into account nothing but the planets themselves, is that they will continue to go round the sun in much their present paths for perhaps a few tens of millions of years, and that within this time nothing fundamental will happen to alter the main features of the present distribution. The theory provides no help at all with the original question of what it was like aeons ago. The calculations simply cannot be done, and there is a fundamental insuperable difficulty here, quite apart from the neglect of all the small unknown bodies and dissipative effects. This difficulty is of a mathematical character, it is of the nature of an uncertainty principle, but involves an uncertainty far greater than that involved in quantum mechanics. It arises from the form of the equations governing the motion, combined with the inescapable feature that we cannot know the positions and velocities (and also the masses) of the planets with perfect numerical accuracy at any single instant. The equations themselves, supposing the planets to be regarded as point masses, are perfectly accurately known in analytical form. They can be set up beyond any question of argument from Newtonian mechanics, but when one wants to solve and to apply these equations, numbers have to be put in at some stage, and as these numbers are related to the positions and velocities none is known quite accurately, or can ever be known accurately. There have been great improvements in the measurements of distances within the solar system in recent years; through radio-astronomy, the distance of the sun from the Earth is now known with an accuracy that is something like a factor of 100 better than it was twenty years ago. Again, passages of space-probes close to other planets are improving the values of their masses (as yet only for Venus and Mars) relative to that of the Earth. But even so, there must always remain uncertainties.

To explain the nature of this curious difficulty, there occur in the solar system what are termed *resonances*. For instance, the orbital period of Venus round the sun is 225 days, while that of the Earth is 365 days; now five times the period of Venus is 1125 days, and three times that of the Earth is 1095 days, and these differ by as little as 30 days, which is quite small compared with either period. When the planetary theory is worked out numerically, a difference like this becomes a small divisor and produces a large term where at first sight only a standard-size term would be expected, and this sort of thing happens in the planetary theory as it is at present conducted on an unlimited scale. If we have two planets, one with period $2\pi/n_1$ days, say, and another with period $2\pi/n_2$, then it is possible to find whole numbers (integers) p and q such that $pn_1 - qn_2$ is smaller than any given number. That is, there are eventually numbers such that

$$| \, pn_1 - qn_2 \, | < \varepsilon \text{ (small)}$$

for any given ε. If one could carry out the procedure of determining the perturbations sufficiently far, sooner or later one would be confronted with divisors of this sort of arbitrary smallness giving rise to very large terms. In practice these usually occur fairly well along in the series, and they are therefore swept under the carpet and forgotten. But if one wanted the theory to predict over very long periods of time, it would not be possible to neglect these singular terms, for they would rise up as it were and destroy the theory. This is how the difficulty presents itself when planetary theory is tackled analytically by means of trigonometrical functions, which is the standard practice. If such a very small divisor occurred really early on, as it does for example in the theory of the four great satellites of Jupiter (here the period of the second is almost exactly twice that of the first, and of the third almost exactly twice that of the second), it would be necessary to proceed entirely differently, and quite different analytical forms would be found to enter.

The nature of the difficulty can be demonstrated fairly readily by considering the possible solutions of an equation of motion leading to resonances. The simplest and most characteristic resonance equation is that associated with the pendulum. The dynamical problem of the pendulum is to predict the angle θ that the rod makes with the vertical: that is, to determine θ as a function of time, $\theta(t)$. The equation of motion governing this is

$$\frac{d^2\theta}{dt^2} + k^2 \sin\theta = 0, \qquad \text{where } k^2 = g/l, \qquad (1.1)$$

and here g is the acceleration of gravity and l the length of the pendulum (which is being regarded as a small heavy mass attached to a rigid massless rod). Suppose the pendulum, which is our analogy for the planetary system, is observed at a certain instant when it is passing through the vertically downward position. Observationally, we are immediately faced with certain inescapable restrictions in that neither g nor l can be perfectly accurately measured, and that the pendulum is not precisely at $\theta = 0$ at the instant of observation. But it is not the fact that k^2 is unknowable with complete numerical accuracy that produces the fundamental difficulty, it is the fact that it is impossible to observe the velocity with which the pendulum passes through the lowest point. For simplicity then, let us regard k^2 as known, and suppose that a velocity of 1 unit is the theoretical speed that would just take the pendulum to the top of the swing. Then if the actual speed is greater than 1 we would have a rotating pendulum, while if the speed is less than 1 then we would have an oscillating pendulum. Now in making actual observations, it is inherent that a series of observations even of the same thing may give different results, and in this case one observer might rate the velocity as 1·01 and another rate it as 0·99, taking these as illustrative figures. Small as this difference is, the consequences are extremely far-reaching, since although they differ only by ± 1 per cent, they would eventually imply a difference in the predicted value of θ larger than any number we care to assign. If more accurate observations were made, such as 1·001 and 0·999, it would merely take a little longer for any arbitrarily large difference in the two predicted values of θ to result.

To see how this comes about, let us consider the analytical form of the solutions for the rotating pendulum and the oscillating pendulum. Equation (1.1) has a first integral

$$\left(\frac{d\theta}{dt}\right)^2 = 2k^2 \cos \theta + C, \qquad (1.2)$$

and three types of motion can occur according as

(*i*) $C > 2k^2$, the rotating case,

(*ii*) $C < 2k^2$, the oscillating case,

(*iii*) $C = 2k^2$, the case when the pendulum just reaches the highest point.

We are concerned here with the rotating and oscillating cases, and taking these in turn, we have:

(*i*) $C > 2k^2$. If we replace the constant C by n, defined by

$$2\pi/n = \int_0^{2\pi} (C + 2k^2 \cos \theta)^{-1/2} d\theta, \qquad (1.3)$$

then with n and ε (which merely defines the origin of t) as arbitrary constants, the complete solution may be shown to be expressible as an infinite series of which the first few terms are

$$\theta = nt + \varepsilon + \frac{k^2}{n^2} \sin (nt + \varepsilon) + \frac{k^4}{8n^4} \sin 2(nt + \varepsilon) + \cdots \quad \text{(I)}$$

This clearly represents a steady rotation at angular rate n given by $\theta = nt + \varepsilon$ on which is superposed an oscillation given by the periodic terms.

(*ii*) $C < 2k^2$. If now we replace C by α, defined by $\cos \alpha = -C/2k^2$, then $\dot\theta$ vanishes when $\theta = \pm \alpha$, and the motion is an oscillation between $-\alpha$ and α where $\alpha < \pi$. The solution can now be represented by another infinite series of which the first two terms are

$$\theta = a \sin (pt + \beta) + \frac{a^3}{192} \sin 3(pt + \beta) + \cdots \quad \text{(II)}$$

where a, β are arbitrary, and p is related to a by

$$p = k\left(1 - \frac{1}{16} a^2 + \cdots\right). \qquad (1.4)$$

Certain interesting things emerge regarding these two forms, for we see that, whereas in the rotating case the solution depends on k^2, in the oscillating case it depends on k. The two solutions are fundamentally different, and it is not possible to pass over continuously from one to the other by any analytical device. In (I) the quantity n gives the mean value of $\dot{\theta}$, and if we regard n as decreasing it will eventually pass through zero to negative values giving a pendulum rotating in the opposite sense. But there is another class of motions, and these are given by (II), for which the mean value of $\dot{\theta}$ is zero.

To see how this would apply to an actual attempt to predict the position of the pendulum: the observer that measured the velocity as 1·01 would clearly be forced to choose the solution (I) to represent the motion, while the observer measuring 0·99 would choose (II). Then the situation would be that, for a certain limited period of time after the instant of measurement, it would not be possible to decide which was the appropriate solution: either solution would supply equally acceptable predictions *for a time*. But eventually, if we continued to observe the system, it would itself reveal whether it was a rotating pendulum or an oscillating one, and thereafter the two solutions would give predictions for θ whose difference would become arbitrarily large. Thus, predictions from the instant of observation can be made only for a limited time, and there is no way *at that stage* of deciding between (I) and (II).

The foregoing is but a simple example of how motion near a resonance introduces this uncertainty. The vanishing of n for the pendulum corresponds in this analogy to an exact resonance $pn_1 - qn_2 = 0$ for two planets, and whenever this happens there is the possibility of two entirely different analytical solutions equally capable of representing the motion, but *only for a limited period of time*. In practice, n_1 and n_2 are themselves not accurately knowable, and what is done in general is to select the standard type of solution proceeding along powers of the masses; but if the resonance occurs early enough it may be necessary to use another solution which may proceed along powers of the square roots of the masses.

Thus, quite apart from the cumulative effect of dissipative action, there is this analytical difficulty permeating the whole of celestial mechanics and making prediction over great lengths of time inherently impossible. If such a problem were posed to a dynamical astronomer with the request for predictions over a given length of time, and one did not supply him with sufficiently accurate constants of integration, he would come back at a certain stage to say that he had reached this kind of impasse and could not proceed without more accurate knowledge of the system.

Nor can the difficulty be surmounted by use of computing machines: a computation started with the velocity 1·01 would lead to a numerical solution completely equivalent to (I), while one started with 0·99 would give numerical values equivalent to (II). The trouble is inherent in the equation of motion because of its nonlinearity, and there is no way of bypassing the difficulty. It is for reasons similar to these that we cannot be sure whether the orbits of the planets have always been nearly circular and nearly in the same plane. For a system such as Saturn's rings, for example, consisting of a swarm of dust particles, it is easy to show that collisions would be so numerous between the particles that they would come to move all in the same plane just because of the collisions. But there can be no guarantee that, for times measured in aeons, the planetary orbits have always maintained their present general configuration: it is not known one way or the other.

The upshot of all this is that in tackling the problem of the origin of the solar system we must first ask ourselves seriously what present features can be regarded as trustworthy clues; and the answer would seem to be very few. It is possible that the total angular momentum is a safe thing to rely upon, but related to this is the consideration that it is not possible to be sure even that a single event produced the planetary system. There can be no certainty, for example, that the highly hydrogenous Jupiter and Saturn have not added to their masses from time to time when the solar system has passed at relatively low speeds through gas and dust clouds in the galaxy, and there is the possibility that on such occasions the sun may from time to time acquire an atmosphere, probably consisting of hydrogen, so extended as to envelop some or all of the planets. It is maintained by a number of astronomers that this situation exists on a small scale even now, with the Earth embedded in a highly tenuous solar atmosphere. But at times both the extent and density of this atmosphere might be far greater. In that event, there would very likely be denser condensations in this solar atmosphere surrounding any planets within it, and there would be opportunities for a massive planet such as Jupiter, sufficiently distant from the sun for the temperature in the solar atmosphere to be low, to gather in hydrogen. At present we have very little knowledge of such processes, but the possibility is there and raises questions to which we would much like to know the answers.

If we cannot be quite certain that the total mass of the planets was originally the same as at present, at any rate in order of magnitude, we may better be able to trust the linear scale of the system, though here again even this is not absolutely certain. The present scale is certainly an important feature that has come about in some way, and so extensive is the system compared with the size of the sun—Jupiter moves at about 1000 solar radii and Neptune at 6000—that many writers have drawn attention to what they regard as an anomaly in that the angular momentum of the planets is so very great compared with that of the sun. The sun has a mass about 700 times that of all the planets put together, but the orbital angular momentum of the planets is nearly 200 times the rotational momentum of the sun. But all this is really saying is that the sun is rotating slowly. If a large mass and a small mass are in orbital motion about each other, it is an inevitable consequence of dynamics that the small mass has the greater angular momentum, in strict inverse proportion with the two masses. So the great angular momentum of the planets follows from their very small masses. It is true that the sun *is* a slowly rotating body, but the solar rotational momentum is an independent property dynamically, until some theory may demonstrate a connexion.

In many theories of the solar system, the sun has been the

prime suspect as the source of the planetary material because it is situated at the centre of the system. But here again, no special significance can be attached to this feature because it is an essential dynamical consequence of the comparatively small masses of the planets. If a number of gravitating bodies are set in motion as a single bound system, and one of them has mass 700 times the rest together, then that one will remain almost stationary in the middle. There is no option about this, and nothing about the origin of the system can really be inferred from the arrangement. However, where the history of the problem is concerned, a number of theories have nevertheless concentrated upon the sun for this alleged reason, and it is of interest to consider some of the difficulties associated with these theories.

To remove material from the surface of the sun, where it is held down by an enormously powerful gravitational field, requires some very violent action. The sun is not readily going to yield up material that it holds at and near its surface, and it is necessary to introduce some comparable force. The obvious possibility is another star, and the English mathematician Jeans devoted much effort to investigating the tidal disturbances that a passing star might produce on the sun. As a result of his calculations it appeared that, if at its closest the centre of the second star came within just over a solar diameter from the centre of the sun, then the gravitational attraction could distort the sun so much that the wave-form travelling across the solar surface would become unstable. This suggested the possibility that a little material would actually detach, rather like drops of spray, from the crest of the tidal wave. Jeans believed that his analysis indicated that there would be drawn from the sun a filament of material that would then segment into a string of planets between the two stars. It is to be remembered that only a small fraction of 1 per cent of the mass of the sun and passing star would need to be removed in such a process to provide the entire mass of the planets. The passing star would meanwhile be moving transversely, and the idea was that its continuing pull on these primitive planets would set them into orbital motion round the sun.

The discussion of the stability of the sun under such forces involves some impressive mathematics, and no doubt when Jeans had completed this limited part of the work he was too exhausted to have much criticism left for other essentials that the theory should meet. A point that was overlooked, not only by him but by many others since, is that when a wave-form travels over the surface of a fluid body this is achieved hardly at all by sideways mass-motion but by up-and-down motions perpendicular to the surface. So the drops of spray would not in fact achieve any serious sideways motion at all as a result of the tidal disturbance, but only possibly by the subsequent transverse pull of the passing star. Another cosmogonist of the day, Jeffreys, criticized this tidal form of the hypothesis on other grounds, for the sun is a slowly rotating body taking about a month to turn round once, whereas the great planets Jupiter and Saturn rotate quickly in periods as short as ten hours. These planets have mean densities not very different from that of the sun yet rotate fifty times faster, and purely tidal forces, which can never generate circulation in a fluid, would seem incapable of accounting for the difference. In an attempt to overcome this, Jeffreys therefore proposed to substitute an actual grazing bodily collision of the sun and star for the merely tidal close encounter that Jeans had studied. The relative speed when the stars slid by each other would be of an order approaching $1000 \, \text{km s}^{-1}$, and this would result in violently turbulent motion in the shallow layer of contact. As the two stars separated, a ribbon of material would be drawn out between them, and turbulent viscosity would endow this with strong vorticity. The internal rotatory motion so imparted would be about directions more or less perpendicular to the general plane of motion of the passing star, thereby possibly explaining why the axes of spin of the planets are mainly perpendicular to their orbital planes. Such calculations as could be made for so intricate a process showed that the general order of the angular velocities thereby produced would be about the same as the planets actually have at present.

But there remains an insuperable difficulty with both the collision and tidal forms of the theory, a difficulty that is in fact even greater with the collision form, and this is that the radial extent of any planetary system so generated could not be very different from the size of the sun itself. It was the American astronomer H. N. Russell who was the first to perceive this difficulty in the 1930s. In those days, when little in the way of computing aids were available, it was possible to give only rough numerical treatments of such questions, but Russell showed that the angular momentum per unit mass about the sun of such a passing star would be far less than that now possessed by any planet, even Mercury, the nearest to the sun, and he demonstrated almost beyond doubt that no process then devised could increase the angular momentum to the extent required. What is supposed to happen in both the tidal and collisional processes is that material is removed from the surface of the sun more or less by a single action. But we know full well nowadays, from experience in launching artificial satellites, that to put anything into orbit the object has to be acted upon twice, or any rate in two different ways. First, the intended satellite has to be lifted clear of the body that it will eventually circulate round, and secondly, this having been achieved, it has to be projected sideways. However powerful a gun one has, it is not possible to put anything into orbit with a single action from the surface of the Earth, for the path must be an ellipse, and the ellipse being symmetrical about its longest axis, which must pass through the Earth's centre, it must again intersect the surface of the Earth somewhere else. A gun could in principle fire an object away to infinity in a hyperbolic orbit, but this would by no means get it into orbit around the Earth.

It is true that the tidal and collision mechanisms do contain a secondary action in the pull of the passing star after the encounter, and it is possible nowadays to investigate the efficacy or otherwise of this force to produce orbiting material.

The paths of small particles moving between the two separating stars can be calculated with a fair degree of rigour by means of modern computers. The action of the stars once the collision is over can safely be regarded as pure attraction by point-masses, and if the small particles are given initially the sideways motions that they would acquire on the collisional form of the theory, this would obviously give them a better chance of getting into orbit than on the tidal theory, but it turns out that the pull of the passing star is utterly incapable of producing the desired result, and particles temporarily removed from the sun would simply fall back onto the surface before getting even half way round. (See Fig. 1.1, p. 21.)

At first sight one might think that by speeding up the passing star sufficiently one could impart any suitable amount of sideways motion, but this is not so: a certain transverse speed would just cause a particle to skim the surface in a circular orbit, but if the actual speed exceeded this by a little over 40 per cent the particle would escape altogether and so not go into orbit at all. Calculations of this kind have recently shown that the difficulty pointed out by Russell is far graver even than he found, and it is practically certain that not so much as a pin-head of material could be put into even grazing orbital motion round the sun, never mind the massive Jupiter at over 1000 solar radii.

Before proceeding, we may pause to mention an alleged criticism of the encounter hypothesis that is still often advanced, yet is entirely baseless. The average separations of adjacent individual stars in the galaxy are so many millions of times greater than their linear dimensions that the probability of a star colliding with another even in a whole aeon of 10^9 years is minutely small. This consideration has been claimed again and again by many writers to rule out any encounter hypothesis. But in fact the concept of 'probability' applies only to the future and not to past events. Scatter a handful of grains of sand on the floor, and the 'probability' that they come to rest where they do when each might have fallen elsewhere is infinitesimally

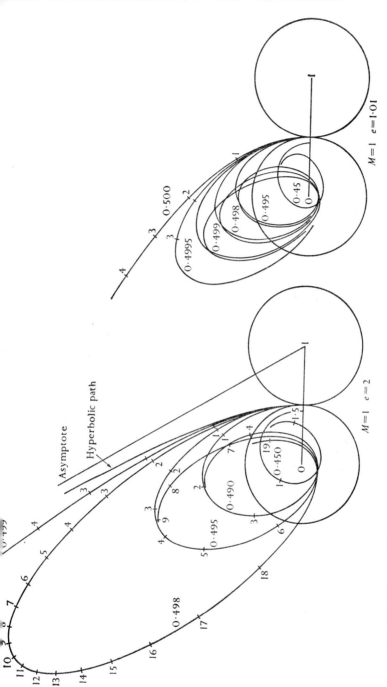

FIG. 1.1 Diagrams showing paths of small particles moving between two separating stars. The numbers, 0·499, etc., attached to the curves indicate the starting-point on the line 01 (regarded as of unit length). The integer markings 1, 2, 3, etc., correspond to equal intervals of time in the motion on each curve.

small. But once something has happened, *probability* no longer comes into it. All that can be inferred from a consideration of this kind is that if the probability of stars colliding in the future is small, then it is unlikely that many solar systems will have been produced by such a mechanism. However, even with the probability minutely small for any selected star, since there are 10^{11} stars in the galaxy, and something like 10^{10} galaxies within the observable universe, it becomes almost a certainty that a collision could happen *somewhere* at *some* time to *some* star in a period of order 10^9 years. But the sun is now by no means a star selected at random: in its nature it is a special star to which planet-formation has happened, and that we are here discussing the problem would require this special *something* to have happened to our sun, no matter how 'improbable' that event may seem when viewed naïvely.

However, returning to the hypothesis of encounter, a simple means of escape from this difficulty of scale was not long in being proposed, assuming the other features of the tidal or collisional mechanism could be relied upon (which, however, now seems doubtful). It was pointed out that there are numerous double stars in the heavens with components similar to the sun, and these are usually fairly widely separated in relative orbits having periods of the order of 10 to 100 years—much the same as the orbital periods of the great outer planets. This means that the separations of such stars would be of planetary orbital dimensions. The hypothesis that the sun may have at one time had a companion star moving at planetary distance would therefore be perfectly permissible scientifically. The main scene of action is now transferred to the companion, and the tidal or collisional mechanism can now be invoked to operate upon it. The picture then is of a third star approaching the system, to make a close encounter with the companion, with the result that a small quantity of material is freed from this companion and the third star. The released material would automatically be at planetary distances, thereby resolving the angular-momentum difficulty of each former mechanism at once.

But it is also necessary to get rid of both the companion and the third star, for there is certainly no massive stellar object at present gravitationally bound to the sun. We cannot simply annihilate them after they have served their purpose, but it is possible so to contrive the three-body encounter that the action of the passing star speeds up the companion during its close passage. It is only necessary to increase the previous orbital speed about the sun by some 40 per cent, which would mean no more than a few kilometres per second, and then the companion and the intruding star would leave at escape speeds in more or less opposite directions, with the Jeans–Jeffreys ribbon-like filament of matter drawn out between them. The velocities of the elements of this filament would automatically be distributed between those of the two stars, and a small part near its centre could have sufficiently low speed relative to the sun—which body meanwhile has stood off at a distance having little or no effect on the violent encounter—to remain behind captured by it. Clearly everything has got to go just right, not only before the encounter but afterwards as well, and considerable objections were again raised on the 'improbability' score when this adjusted form of the hypothesis was proposed; but, as we have seen, so long as a set of events *can* happen it is legitimate to consider them as a possible explanation. There have been many instances of golf-balls colliding in mid-air, and there is even on record an occasion when after colliding both balls fell into the hole. But once such a thing has *happened*, it is no use offering disbelief on the grounds that it is too 'improbable', though long odds may safely be offered against such an event occurring on any specific future occasion.

So infinitesimal is the probability of such an event as a stellar collision that at one time on the basis of this type of theory it appeared that the solar system might even be unique in the galaxy. This possibility ran counter to the widespread uncritical assumption of many that the world is here for the purpose of providing an arena for man's activities. The idea that the Earth's existence might depend on mere chance greatly

3—M.S.S.

disturbs the holders of such views, with the result that they will snatch at any argument, valid or not, to try to rid science, as they conceive it, of an emotionally distasteful hypothesis.

Where the theory could more properly be criticized was over the question of whether material suddenly removed from a star could immediately condense into planets. Almost all the material of a star is at a temperature well in excess of a million degrees, and if a mass equal to that of the planets is considered to be localized in and below an area of a star, most of it will be at a temperature of this order. In a collision process, even further heating would result from the energy of relative motion. So when the stellar material was released, it would find itself more or less free in space at this very high temperature; and though it has never been proved there is considerable doubt whether it could hold together at all. The planets themselves have masses much too small to control thermal speeds of gases corresponding to such temperatures; so there would be a strong tendency for the material to expand and thereby cool. The actual situation would be complicated by the close proximity of the two stars, which even an hour after the collision would have separated only by a radius. As the material expanded, some of its thermal energy would be used in doing work against the attractions of the stars, and this means against a gravitational field hundreds of times stronger than the pitifully weak planetary attraction. It would therefore be wrong to conclude that the hot material would necessarily disperse in a matter of minutes.

Another question, but again of an inconclusive kind, is whether enough material could be dislodged from the surface of the colliding stars. So strong is the gravitational field there that a great deal of energy is needed, and unless the whole of the ribbon of material were captured by the sun, so that the process was extremely efficient, a high initial relative velocity of the two stars would be needed. It seems more likely that only a central part of the filament would be left behind held by the sun, and that the two ends would simply fall back to the surface of the adjacent stars; if so, then a high initial relative velocity of the stars would be needed, though not an impossibly high one judged by stellar standards.

Yet another variant of the encounter theory was to postulate that the companion of the sun was itself a close double-star with components almost in contact. Here again, examples of such triple systems are known in the heavens; so this also would be a permissible starting-point. Addition of matter to the double-star companion would occur if the whole system passed through gaseous nebulae (which latter are extremely numerous), and this could result in the two components eventually merging together into a single rotationally unstable star. This would then break up and there would be an opportunity for material to be showered out into space; in addition, the recoil on the resulting single companion might be enough to sever its weak gravitational bond with the sun.

At this point of the history of the problem, Hoyle proposed yet another variant of the single companion-star idea. The motivation of his proposal was that stars such as the sun, and a great many others besides, were believed to be composed almost entirely of hydrogen and helium, only a few per cent by mass consisting of other elements. But some of the planets are very rich in heavy elements, such as oxygen, silicon, and perhaps iron, and a source able to provide these in suitable proportion seemed to be an essential requirement. Thus the suggestion made was that the companion might have been a star much more massive than the sun, perhaps twenty or thirty times as great, which reached the supernova stage and exploded. When a star has consumed its hydrogen and helium, it collapses rapidly, thereby increasing its central temperature sufficiently to enable heavy elements to be synthesized. A second consequence of the collapse is a great increase in the speed of rotation of the star, and this may produce rotational instability. This results in material being thrown off the surface, now in vast amounts. The supernova process is actually observed to take place occasionally for individual stars, and it has been estimated that a quantity of material of the order of the mass of the sun can be ejected into space. If that is correct, then the sun would only have had to catch the tiniest wisp of the whole amount to capture something of total planetary mass.

The actual ejection would occur rather like a catherine-wheel, but there would be no reason to expect perfect symmetry, and the slightest recoil on the companion resulting from asymmetry would speed up the residual star by the trifling amount needed to send it away from the sun. The speeds of ejection of the gas are hundreds of kilometres per second, and most of the material, even after escaping from the strong field of the companion, would still have enough energy from the expansion to escape also from the much more distant sun. Ejection would be in almost all directions at a wide range of speeds, but some minute proportion—much less than 1 per cent is all that is necessary—could have insufficient speed to escape from the sun. This would be the crucial wisp of incipient planetary material. But on this mechanism there is no possibility of any immediate condensation into planets, and instead the material would go to form a cloud of gas and dust round the sun. Angular momentum would again result of the same order per unit mass as that originally possessed by the companion. The general expansion away from the massive companion would cause the material to cool, and then after capture by the sun further cooling would occur in such a way as to keep the material more or less in equilibrium with the solar radiation it was receiving. This would mean quite low temperatures, of the order of a few hundred degrees absolute for material at the distances of the planets. The cooling would mean that solid and liquid particles would condense out from the original material, but elements such as hydrogen and helium would remain in gaseous form. The question then is whether the planets can form subsequently within this gas and dust nebula surrounding the sun.

As a result of much recent work, it has come to be widely supposed that any theory of the origin of the planets should begin by finding some mechanism for endowing the sun with a nebula of this kind as a preliminary, though there remain great difficulties in showing, with anything approaching rigour, how planets could form within the nebula. But if a combined gas and dust cloud is regarded as circulating round the sun, enveloping it in perhaps a lenticular-shaped nebula, an interesting development would occur in that the dust will soon come to settle into a thin flat disk resembling a gigantic Saturn's ring, though situated much farther out in proportion to the size of the sun. The reasons for this are simple. For a heated gas-cloud rotating round the sun, there can be a pressure-gradient perpendicularly away from the general equatorial plane of the distribution, and this can maintain elements of the material moving in circles round the general axis of rotation and always at the same distance above, and below, this equatorial plane. But such gas-pressure would not act on dust particles, and apart from minute effects of gas-resistance they would begin by moving in paths that must automatically lie in planes through the centre of the sun. At first these planes might be distributed in more or less all positions, which would mean that the dust particles would collide; and such collisions would cease only when all the dust had come to move practically in one plane. This is why Saturn's ring is so thin—even where it is not transparent it may be only one particle thick—for if any particle should get out of the plane of the ring, its subsequent path would cause it to cross through the plane of the ring twice per revolution, and sooner or later a collision would occur that would tend to bring the particle back into the general plane of the ring. There would still be minor lateral collisions between particles not moving in perfectly circular paths, through perturbations or other causes, and as these effects can never finally be got rid of there would result a gradual evolution of the ring. Accordingly, soon after its formation, the solar nebula must probably be pictured as consisting of a more or less axially symmetrical gas-cloud, of radial extent from the sun possibly equal to that of the present orbit of Neptune (which means about thirty times the Earth's distance) and a depth perhaps of about a tenth of this, giving the whole thing a lens-shaped form. In addition, within this at its equatorial plane, there would be developed this disk of dust, as thin as possible consistent with the total amount of matter in it. It is possible that the total mass of the dust would be only a few per cent at most of that of the whole nebula.

Now there are several ways in which the sun might come to acquire such a gas and dust nebula. For instance, as earlier mentioned, such material already exists in abundance in galactic space, and the question arises whether the sun might not simply acquire enough such material to produce the planets on going through a suitable interstellar cloud. There may well be more than one way of effecting this; certainly a number of mechanisms have been proposed. One relies on a process related to a little-known controversy in dynamics that has been satisfactorily resolved only in recent years. The problem was this: If we have a pair of stars or point-masses bound together in orbital motion, can the arrangement be disrupted through purely dynamical attraction by firing in a third mass suitably from infinity, and thereafter the three stars disperse to infinity in different directions? In fact, it was over the reverse form of the question that uncertainty reigned for many years, namely, can three attracting point-masses come together from infinity in such a way that two of them are left bound together in a mutual binary orbit and the third leaves the scene? The problem was argued theoretically at intervals for some decades, and different opinions maintained, but the matter was not satisfactorily resolved until the Russian mathematician, Sizora, actually worked out a case in which the capture occurred. He discovered a set of initial conditions of projection such that when the motion was followed out numerically, the stars interacted (without bodily collision) in such a way that a double-star was left behind formed of two of them. This ended the mild controversy.

Another Russian, the astronomer Schmidt, took up this idea and developed it for cosmogonical purposes, replacing one of the stars by a cloud of gas and dust. One of the two stars would of course be the sun, and the more or less simultaneous encounter of the sun and cloud with another star representing the third body could obviously result in part of the cloud being captured by the sun. There is nothing highly 'improbable' about such an event, for galactic clouds have dimensions comparable with interstellar distances, and gravitation acts equally well, only with more leisurely speeds of the bodies involved, at large distances as at small ones; so the passing star does not have to come especially close to the sun. Nor need the sun capture much of the cloud, which might in its entirety have mass of stellar order. So this is a way that stars might very frequently come to possess clouds of gas and dust moving round them. In any particular case, the amount of material captured and the associated angular momentum, which is what would eventually determine the scale of the system, would depend on chance factors; but manifestly there would be no difficulty in accounting for the scale of the actual planetary system—unless, indeed, it may now appear even rather on the small side to be consistent with such a process.

There is an even more direct way in which a star can collect material, without the intervention of a second star. This is by a process of accretion through its own gravitational focusing as it passes through an interstellar cloud. If at one extreme we imagine a star simply at rest in such a cloud, its gravitational attraction would pull material radially inwards. If there were no rotational momentum associated with the cloud, it would all eventually collapse on to the surface of the star. But if the cloud had some rotational momentum, this would be strictly conserved, and the cloud could only contract so far to a stage in which this situation enabled centrifugal force aided by gas-pressure to balance the attraction of the star. The mass of such a cloud might be of the order of the sun itself, and this would be far more than would be needed in order to make the planets. At the other extreme, if the star passed through the cloud at very high speed, say some hundreds of kilometres per second, the star would collect practically only the material that it ran bodily into, and in total mass this would be quite negligible. Somewhere between these extremes there would be a relative speed of star and cloud that would lead to a capture of mass equal to the total planetary mass.

When the star moves through the cloud there is of course a gravitational interaction. If we imagine the star at rest with the gas and dust streaming by it, the pull of the star focuses the material into the axial line behind it. The closer the material passes to the star, the more strongly will it be pulled in and the more its path curved towards this line, whereas for material moving sufficiently far out almost no such curving would happen. The convergence towards the axis results in opposing transverse streams, and the resulting loss of this sideways component of the motion, through its conversion to random energy, enables the star to capture the material. The distance out to which it effects this is greater the smaller the relative speed of cloud and star, and it can be shown that the distance is of the order of $2GM/V^2$, where G is the constant of gravitation, M is the mass of the star, and V the relative speed. With this result available, it is then a simple matter to calculate in order of magnitude the total amount of material that a star would capture in passing through a cloud of any given size and density.

The densities in the clouds are extremely low by ordinary standards. The gas has a density possibly in the range of 10^{-22} to 10^{-23} g cm^{-3}, and the dust usually associated with such clouds may contribute about 1 per cent of the whole mass. The gas is believed to be mainly hydrogen, but there is no reason why helium or other gases may not also be present to a small extent. These clouds as observed are extremely irregular in size, shape, and distribution, but their dimensions can be said to be of the order of parsecs (1 parsec $\simeq 3 \times 10^{18}$ cm), though a great variety of sizes occurs. If we imagine a star such as the sun passing through one of these clouds along a track a few parsecs in length, in order for it to gather in material equal to the mass of the planets the relative speed would have to be about 0.5 km s^{-1}. The total time taken for passage through the cloud at such a speed would be a few million years.

A further feature of these clouds is that they are generally in slow rotation with an angular velocity of order 10^{-15} s^{-1}. This would mean that in a million years the clouds would have rotated only a few degrees or so—not enough seriously to complicate the process of capture. Nevertheless, even this minutely small rotation implies considerable original angular momentum for the material captured, and it happens that this is about of the same order as that possessed by the planets. But this may not be the only source of angular momentum: the clouds are quite irregular in form and distribution, and also in density, especially on the large scale, and they may have small internal velocity differences. The amounts of material converging to the accretion-axis from opposite sides of any plane through it might therefore not be perfectly equal, and the slightest asymmetry, as small as 1/100, in the amounts of material arriving from opposite sides would also be adequate to account for the required angular momentum. This point is of some importance, for it is possible that the general rotations of the clouds are strongly related to the orbital rotations in the galaxy, whereas the invariable plane of the solar system, defined by the total angular momentum of the planets, is some 60° inclined to the present galactic plane. The interstellar cloud concerned here would of course have to have been in existence 5×10^9 years ago, and we can scarcely hazard even a guess as to the extent to which the state of the galaxy at that time may have resembled the present state.

However, there seems little doubt that the sun could have acquired an enveloping gas and dust nebula of suitable mass and angular momentum by this unaided action of itself on interstellar material. The average speed of the sun relative to stars in its near neighbourhood is about 19 km s^{-1}. The relative velocities of the stars arise from slight differences in the much higher individual speeds that the stars have in their galactic motions. All stars in the neighbourhood of the sun have speeds of the order of 250 km s^{-1} which carry them in more or less circular paths round the galaxy. But just as the planetary orbits are not quite circular and not quite in one plane, so also are stellar orbits in the galaxy. The result is that the stars in any comparatively small local region of the galaxy have slightly different motions, and as viewed from any one, such as the sun, appear to have relative velocities of the order of 10 to 30 km s^{-1}. And the same must hold for the velocities of the stars relative to the interstellar gas-clouds, for the latter must also be in gravitational orbits in the galaxy. For a whole group of individual stars, these relative speeds are distributed rather like the maxwellian distribution of velocities for the particles of a gas, the probability of very large or very small velocities diminishing rapidly. It is possible by analogy to estimate roughly the probability of a star having relative velocity with respect to the clouds of 0·5 km s^{-1} or less, and this is found to be of the order of 10^{-4}.

In its whole history the sun, or any other star with an age of 5×10^9 years or more (and some stars are now believed to have ages four or five times this), must certainly have passed through some hundreds if not thousands of interstellar dust clouds. As we shall see in Chapter 5, the existence of huge numbers of comets in the solar system probably provides direct evidence of this. Since the sun may have existed for several aeons before the formation of the planets, a slow encounter of the kind needed would in any case have prior probability of order 10^{-2}, but as we have seen there is no need to rely on such a consideration for a particular case, such as the sun with its retinue of planets represents, for so long as the requisite low speed *can* occur then it may legitimately be adopted as a postulate of a theory. On the other hand, it would follow that a proportion, of order 10^{-2}, of all old stars will at some time have undergone an encounter with a cloud at sufficiently slow speed to have brought about capture of an amount of material of planetary order, and this might mean that 10^8 stars in the galaxy have at some time experienced this process and therefore gone on to possess planets. The number might even be slightly higher than this, for the process of accretion will usually involve also a braking action on the star, reducing its velocity relative to the cloud, so that an initial speed in excess of $0 \cdot 5$ km s^{-1} might still be adequate. The effect of this would clearly be some increase in the foregoing estimate of the frequency of suitable encounters.

Yet another means whereby the sun might acquire such a disk of material of planetary extent has been proposed by Hoyle as a result of his conclusions regarding the formation of stars. In brief, stars are regarded as forming by Helmholtz-contraction (which means purely gravitational collapse resisted only by gas-pressure resulting from compression temperatures) of condensations within a far larger self-gravitating galactic gas-cloud. Such a large cloud, formed possibly within a spiral arm of a galaxy, would inevitably have only very slow rotation, but on collapsing extremely rapid rotations might develop as a result of conservation of angular momentum. Many stars, especially those with high mass, are found to be still rotating very rapidly, but not stars such as the sun. As a rotating gaseous mass contracts in size, instability may take the form of a sharp edge developing at its equatorial circumference, resembling the shape of a lens. Further contraction will result in material spilling out through this edge into regions closely surrounding the equatorial plane. The time of rotation of the sun at the stage when this would begin to happen could be only a few hours, and become even shorter as the contraction proceeded. But the material ejected into the equatorial plane would have practically the free orbital period appropriate to its distance, and with increasing distance this would be longer than the rotation period of the sun. Thus the sun would be spinning faster than its surrounding disk of material. If ordinary material contact were maintained between the sun and disk, viscosity would tend to equalize local rotation rates, and angular momentum would thereby be transferred outward from the sun to the disk. But the actual amount of this would be negligible, and to bring about a really important degree of transfer the theory postulates a very strong magnetic field for the sun at this stage of its existence, brought about as a result of a pre-existing extremely weak field permeating the primordial cloud from which the sun condensed. The presence of the magnetic field would more or less connect the disk to the body of the sun by lines of force, rather like spokes of a wheel, though here the 'magnetic spokes' would stretch in an elastic way, and would urge the material forward in its path in an endeavour to equalize the angular velocity with that of the sun. But the main controlling force would always be the solar gravitational attraction, and the actual result of the magnetic coupling would be to transfer angular momentum to the material of the disk causing its various parts to spiral outward farther and farther. In this way, almost all the original angular momentum of the sun would be transferred to the disk, and clearly if this were high enough and the process continued undisturbed the disk would ultimately attain planetary dimensions. It could then cool to temperatures appropriate to the distance of the sun, and a disk of gas and dust result.

The theory rests strongly on other theoretical realms that are still largely speculative, and it may be some time before its postulates can be verified or disproved. The required magnetic field is thousands of times greater than the present general field of the sun, and whether the process of magnetic coupling could really work as effectively as is required has not been conclusively shown. Moreover, that stars have condensed from a state of far greater extension, any given star having achieved its individuality from the moment of fragmentation of a far larger cloud, is open to some doubt in view of the existence of large numbers of close double-stars, for if these are imagined to be expanded again there would soon simply not be room for the two components with the present orbital separation. This means that formation by contraction would have to be supplemented by some subsequent process reducing the separations, and the only mechanism that seems capable of this without increasing the angular momentum per unit mass is large-scale accretion. But if this can substantially change the masses of the stars, there can be no essential reason for supposing that the whole mass of a star such as the sun has been primitive to it, and that the whole star was formed in a single process of more or less spherically symmetrical contraction. Present observational evidence suggests that stars similar to the sun rotate slowly, and it leads to difficulty to suppose that all stars must originally have rotated rapidly and that only certain classes have been subsequently slowed down.

As has been said, almost all recent attempts on the problem of the origin of the planets have regarded the production of a nebula of gas and dust as the prime requisite. But even if a thoroughly satisfactory theory of this existed, there would still be the very large question of how or even whether the nebula could subsequently evolve into a planetary system. The initial evolution of the dust, in so far as its motion is concerned, does not seem difficult to predict, and as we have seen it must almost certainly come to form a thin planar disk of material. On the other hand the gas, heated by the sun, will have three-dimensional extent, and possibly take up a flattened lens-shaped form. This difference between the gas and the dust distributions may well be of the highest importance if planets are to grow, for if a completely uniform volume-distribution subsisted throughout the nebula, with the density much the same everywhere, the action of the sun would probably prevent any accumulations forming at all, for the following reason.

If within such a cloud a condensation is to occur, the tendency of an element to pull itself together must exceed the power of the sun to pull it apart. For a small element of linear dimensions r situated at distance R from the sun, the part nearer to the sun will be more strongly pulled than the remotest part by an amount $2Mr/R^3$, while the element itself will exert a self-attraction at its surface of $\frac{4}{3}\pi\sigma r^3/r^2 = \frac{4}{3}\pi\sigma r$, where σ is the density of the element. On this simple analysis of the situation, the self-attraction will predominate if

$$\tfrac{4}{3}\pi\sigma r > 2Mr/R^3. \tag{1.5}$$

Now if we suppose the sun expanded until it just reached the element we are concerned with, the reduced density ρ of the sun would be given by $M = \frac{4}{3}\pi R^3 \rho$, and the above condition becomes

$$\sigma > 2\rho. \qquad (1.6)$$

This gives a simple rule-of-thumb means of settling whether condensations can form despite the tidal action of the sun, namely that they can do so only if the local density σ exceeds about twice the mean density that the sun would have if its radius extended to the region concerned.

Application of this rule shows at once that the planets could not be expected to form in an entirely gaseous lenticular nebula surrounding the sun, for even if such a distribution had thickness only about a tenth of its radial extent, the overall density could only be about a hundredth of that that the sun would have if expanded in this way, for the whole mass of the planets is only just over a thousandth the mass of the sun. This conclusion would not only relate to condensations in the outermost part: nearer the centre the inequality would be even less satisfied.

But the same does *not* hold for the plane disk of dust, because for this the dimension perpendicular to its plane would be minutely small, and the volume density of the dust-disk would be of the order of unity, since most solid materials have densities of this order. This is much the same as the mean density of the sun before any imagined expansion of it, namely $1\cdot4$ g cm^{-3}. Accordingly, self-gravitation within the disk can far exceed the solar disruptive effect. For this reason it seems to be an essential requirement for planetary development that there should be such a disk of dust, but the detailed working out in adequate mathematical terms of exactly how the disk would evolve and the time-scale associated with the growth of planetary masses remains to be accomplished at a satisfactory level.

To begin with the particles of the disk would all be in orbital motion round the sun, so that the various parts would have greater speed and greater orbital angular velocity the nearer they were to the sun. But as has been seen the density would be high enough for condensations within the dust to grow, and the probable course of development would be the formation of numerous aggregations at many parts of the disk. The power of any such condensation to grow as compared with another one is proportional at least to its mass and size. Larger ones could therefore outrun smaller ones in gathering in material, and even perhaps be capable of gathering in smaller ones. In this way a few final condensations might form before all the material was collected in, and it is possible, though it has by no means ever been proved, that a comparatively small number of primitive planetary bodies at first developed from the disk of dust. If any pigmy incipient planets were left with no further material to acquire, they might be capable of being associated with some of the known asteroids, some of which have diameters measuring a few hundred kilometres. If such a description of the process is correct, then it would evidently yield a set of planets moving in the original plane of the disk, and it would also indicate that all planets would begin as objects formed from the dust cloud. In that event, it may well be that deep down in the centre even of planets such as Jupiter and Saturn there is original material of the dust cloud from which the planet first began as a condensation in solid form.

Even among the terrestrial planets there is a considerable range of masses: the Earth exceeds Venus in mass only by about 20 per cent, but is about nine times as massive as Mars and twenty times as massive as Mercury, and with the moon perhaps one of these original primitive planets, the Earth exceeds it about eighty-fold. But the status of the moon is by no means settled, and it is an open question whether it developed in some way through instability of the Earth, or whether it was subsequently captured from an independent orbit. It may be conjectured that the largest condensation forming in any given neighbourhood would dominate the scene locally and, besides sweeping up all the dust, be able to capture small condensations. As possible evidence of this it seems highly likely from the nature of their orbits that the small outermost satellites of Jupiter and Saturn are comparatively recently captured asteroids, but whether the great Galilean satellites of Jupiter, which so closely resemble our own moon, at least in mass, represent bodies captured early on is quite unknown.

It is to some extent necessary to account for the present rotations of the planets, though much may have happened to alter these over the whole age of the system. However, there is no intrinsic difficulty on the dust-disk hypothesis in accounting for rotation periods of the order of a few hours. In virtue of the general orbital motion round the sun, the material of the disk will possess vorticity, just as if it were a fluid. It is a well-known theorem that if a small element of volume of a fluid could suddenly be solidified, it would thereafter rotate with an angular velocity roughly equal to the local value of the vorticity, the precise rate depending upon the shape of the element solidified. In the present case, despite the fact that the inner parts of the disk are travelling faster than the outer parts, the vorticity is in the same sense as the angular motion of the disk; this means that the condensations would rotate in the same sense as that of their angular motions round the sun, as is more or less the case for the Earth at the present time. This process would be capable of producing rotation periods as short as a few hours, depending upon the range from which the condensations were able to gather in the material of the disk. But it is still highly uncertain what original rotation rates have actually to be accounted for. Both Mercury and Venus are believed to rotate extremely slowly, and if they have now reached some steady condition it will be a matter of great difficulty to determine their original state. The situation may not be so hopeless in regard to the Earth because of our special facilities for studying it, and because of the presence of the moon, which could have strongly influenced the rotation of the Earth, and may still be doing so; but it would be necessary to know for what period the Earth–moon system has been isolated and to what extent autonomous. Mars, with a rotation period of just over $24\frac{1}{2}$ hours and an obliquity of $25°$, is very similar to the Earth but has no massive moon to affect it and has smaller mean density.

The latest radio measurements of Venus not only confirm the slow rotation that optical astronomy had long since assigned to the planet, but surprisingly indicate that it is in fact retrograde with period about 250 days. How a planet so similar in size to the Earth yet without any sizable satellite could have come to lose a comparable quantity of rotational momentum is something of a puzzle. It is known that solar tides would be considerably stronger on Venus, but whether they would be adequate for this purpose cannot be decided until more is known of the internal structure of that planet. One suggestion that might explain the past occurrence of considerable braking action is that Mercury was once a satellite of Venus. This would mean a satellite three or four times more massive than our moon, and if it were driven outwards by tidal frictional action it could reduce the rotation period to a matter of a month or more before it would itself escape to become an independent planet. The subsequent free orbit of such an escaped satellite would at first pass near that of Venus itself, and considerable evolution would be needed to alter it to the present orbit of Mercury, but as we have seen such changes might just possibly happen through purely dynamical action, or as a result of dissipative effects, over a period of time of the order of aeons.

Difficult as it has proved, and is still proving, to account for the planets themselves, the presence of their numerous satellites must also be explained before we can regard any theory of the solar system as completely satisfactory. Many attempts have been made to devise some process whereby the moon might originate from the Earth through some instability of the combined body as a single planet, but it now seems dynamically impossible that such an event could occur. Also, any such process would be of little help in explaining the satellites of the four great outer planets, for the masses of these bodies are minute compared with their primaries, and their orbital rotational momentum is negligible compared with the rotational momentum of the controlling planets. One suggestion for producing these has been that, rather as a disk of dust forms round

4—M.S.S.

the sun, so round any planetary condensation there would form
a localized disk but now in rotation round the planet, and that
this disk would condense into objects in orbital motion about
the central planet. As for the development of the solar disk,
mathematical proof that such a process would take place has not
been forthcoming, doubtless because of the extreme difficulty
of adequate theoretical treatment of such problems. However,
the abundance of close satellite systems of the great planets
certainly suggests that satellite formation must more or less
inevitably follow upon planetary formation. If such a process
happened to the Earth, it might by analogy be expected that a
number of small moons would result, and then some subsequent
collection of these into a single moon would need to have
happened. For entirely different reasons, such an earlier state of
the moon has recently been proposed by several astronomers,
who believe that the present rate of recession of the moon is so
great that the moon would have been skimming the surface of
the Earth less than a single aeon ago! The tidal forces responsible for this depend on the square of the mass of the satellite;
if the moon were divided into four, say, the tidal lifetime could
be extended fourfold to something more acceptable.

Yet other suggestions are possible, however, for the formation
of the satellites. As a planet grows within a disk of dust and gas,
the distance to which its gravitational pull can reach in contest
with the ever-present central sun increases, and the amount of
angular momentum per unit mass brought in will also increase.
Now there is a limit to the amount of angular momentum that
can be stored in any given gravitating body, and if that limit is
exceeded the body will become rotationally unstable and break
up. It might be that the amount of material in the original disk
was insufficient for this limit ever to be reached by any planet,
but on the other hand the general question would remain as to
how a planet would arrive at a stable form if the amount of
material brought in did produce rotational instability. For very
many years it had been thought that the result of such rotational
instability on a body of more or less uniform density, such as an

initially formed planet, would be a break-up into two roughly equal masses in close orbital motion round each other. Even the numerous close double stars in the sky were thought at one time to have originated in this way despite the fact that their components are gaseous and highly condensed towards their centres. But modern investigations of the problem have disclosed that, in order to satisfy the requirements of energy and angular momentum, if break-up into two main pieces results, these must have an intermediate mass-ratio of something like 10:1. Equality is quite impossible, and so also would be a large ratio such as 100:1. Moreover, the two pieces must separate completely in hyperbolic relative motion.

If such a catastrophic break-up of a rotating planet occurred, there is always the possibility of a stream of minor droplets forming between the two main separating bodies, after the manner of the Jeans–Jeffreys encounter mechanism. Some of these droplets might escape from both the main masses, while some adjacent ones might be captured by them, and it can be shown that there is a good chance of some going into orbital motion round the larger mass, but not round the smaller one. On the basis of these considerations, the possibility has been pointed out that the whole of the terrestrial group could have resulted from rotational break-up of a single large primitive planet of which Jupiter, say, is now the surviving larger component. It must be remembered that the total mass of the four inner planets is less than 1 per cent that of Jupiter now, so they would scarcely amount to more than mere droplets in such a system. The similarity of the moon to the four large satellites of Jupiter has already been noted. The initial orbits of planets so produced could not of course be circular, but subsequent dissipative action or even purely dynamical action might be capable of rounding them up in several aeons of time and bringing them to their present positions. As for the smaller mass thrown out of Jupiter, the surface escape-speeds for all four great planets are far higher than their orbital speeds, and hence than the escape-speeds from the sun at their respective distances. Accordingly the ejected smaller portions without satellites could be thrown right out of the solar system, and this would account for the absence of any planet with mass about a tenth that of Jupiter and possessing no satellites.

If, however, it is the case that some at least of the inner planets developed originally within the disk of dust, the same question arises about their means of reaching rotational stability. It happens that the Earth and Mars can be paired together satisfactorily as components resulting from a single rotationally unstable planet since their mass-ratio is as high as 9:1. The same may hold for Mercury and Venus as a pair. The surface escape-speeds from the inner planets are low compared with their orbital speeds, just the reverse of the great outer planets, so that anything projected from the Earth could not be expected to escape from the sun, but would simply become an independent planet. On this basis, the moon could be regarded as a droplet formed between the Earth and Mars after the latter was spun off, and as the moon has mass only just over 1 per cent of the combined mass of the Earth and Mars there is nothing excessive in regarding the moon as a mere droplet from the whole system. Also, consistently with the dynamics of the process, the moon could remain a satellite of the larger component, but the smaller component, Mars, could not thereby obtain any satellites. This is not really inconsistent with the possession by Mars of its two satellites, Phobos and Deimos, each no more than a few kilometres in diameter, for on the score of their very smallness and proximity to the planet they seem not to be genuine full-scale satellites but much more likely to represent subsequently captured asteroids, though the precise mechanism by which this may have happened has yet to be settled.

The hypothesis of initial growth from a disk of dust leads on to a reasonable explanation of the significantly smaller densities of the outer planets, which must have considerable content of the lightest gases. Jupiter has an average density of 1.33 g cm^{-3}, that of Saturn is as low as 0.69 g cm^{-3}, the smaller Uranus has the value 1.56 g cm^{-3}, and Neptune 2.27 g cm^{-3}.

Since all these bodies are far more massive than the Earth, which has mean density 5.52 g cm^{-3}, it is plain that they cannot possibly have more than a small portion of their mass made of terrestrial-type material. But there is no serious difficulty here, for the Earth and other terrestrial planets could not even now retain hydrogen atmospheres, in brief because the thermal speeds of hydrogen atoms and molecules at these distances from the sun are too high to be controlled by the small gravity-fields of the terrestrial planets. On the other hand, if the Earth were transported out to the distance of Jupiter or Saturn, the temperature maintained there by the sun would be sufficiently low for hydrogen to be captured and retained by the Earth. Thus if the solar nebula consisted of both a disk of dust and a surrounding cloud of light gases, these primitive planets forming from the dust in the outer parts of the system could thereafter gather in light gases. It might seem from this that the largest and least dense planets should be at the extreme outside, but there is the additional factor that the power of gases, through solar heating, to escape from the sun itself falls off less slowly than the gravitational power of the sun to retain them, and therefore it could be expected that there was less of the solar nebula at the distances of Uranus and Neptune than in the neighbourhood of Jupiter and Saturn.

Just as the material of the dust-disk would possess vorticity that would eventually manifest itself as the planetary rotations, so also would the gas of the solar nebula; and as this was drawn in on to the primitive planetary nuclei it could add angular momentum of sufficient strength to account for the rapid rotations of the great outer planets. But where these rotations are concerned, the axis of Uranus exhibits a striking anomaly in that it is tilted at $98°$ from the upright direction that would be expected from the position of its orbital plane. The spin is rapid, with a period of less than 11 hours, but the actual inclination means that on the whole the planet is rotating slightly backwards! This is certainly not explicable as the result of growth of a single body in the original solar nebula, and the only possible explanation would seem to be that at some stage two separate comparable planets came together and coalesced into a single mass. If this happened, the final rotational vector would be the sum not only of the two original rotational vectors but also of that associated with the relative orbit in which they came together. With the slightest inclination of the planes of the two orbits (round the sun), the plane of their close relative path could be at almost any angle to either orbital plane, and the final result would be rotation about an axis in almost any direction, quite minor circumstances determining precisely where it would come. Clearly the satellites of Uranus would have to be a later development, but there is no particular difficulty here, for if two fairly equal planets joined together in such a way, the combined body would almost certainly be rotationally unstable, and fission of it would occur in the manner already described with the possibility of producing satellites.

There is no easy way to conclude an essay on this subject, for as we have seen the whole problem is very far from concluded, and even if the general hypotheses that we have discussed above are correct the working out of them to show what processes result can scarcely be said to have been more than begun. But the picture that seems to emerge at the present time is almost a complete reversal from that proposed half a century ago: then the general idea was of immensely hot planets wrenched violently out of the sun and left subsequently to cool, whereas now the picture is of cool material from other stars, or perhaps captured from interstellar clouds, gradually accumulating into small solid planets, and then at the outer parts of the solar system carrying on to become composed mainly of lighter gases, though not necessarily in gaseous form. If this latter description is generally correct, some other cause of subsequent evolution of planets than a mere cooling down, at any rate of our own Earth, will be required. This important matter will be taken up in the next chapter.

2 *The Interior of the Earth*

IT was remarked by Eddington many years ago that practically the whole of the material of the universe is tucked away and for ever hidden from us behind inpenetrable barriers. Eddington was thinking principally of the stars, which are only observed by light that comes from thin rarefied layers at their outer surfaces. It is this that makes astronomy so essentially a theoretical science, for it is only by theoretical means that the properties of the invisible parts of the universe can be inferred from the light and other signals emanating from a minuscule proportion of it. But it is seldom adequately realized that much the same holds in varying degrees for the planets, including even our own Earth. The deepest borings yet made measure but a few kilometres, and these at only comparatively few points: yet the radius of the Earth is some 6371 kilometres and its surface area in excess of 500 million square kilometres, so that even if such borings could be made everywhere, and the material encountered brought to the surface for examination, it would mean that less than a thousandth part of the mass of the Earth would be available for this closer inspection. In fact, of course, the amount accessible to direct study, supposing that to be meaningful philosophically, is far less even than this, and it is doubtful if as much as 0·01 per cent of the Earth could be so studied; the properties of the remaining 99·99 per cent can be known at present only theoretically, and it is the thesis of this essay that it is the behaviour and properties of this hidden

99·99 per cent that mainly control the superficial 0·01 per cent
that we can see.

From the point of view of the problem of the origin of the
planets, it would clearly be of the greatest interest to know
what substances they are made of, or, failing such detailed
information, at least whether they are made all of the same
material. The mean densities of the planets make it plain that
certainly the great outer planets must be largely made of quite
different substances from those of the terrestrial group and the
moon. The masses and densities are given in Table 2.1.

Table 2.1

Terrestrial planets			Great outer planets		
	Mass (Earth = 1)	Mean density (g cm^{-3})		Mass (Earth = 1)	Mean density (g cm^{-3})
Mercury	0·054 (± 0·005)	5·0 (± 1·2)	Jupiter	317·8	1·33
Venus	0·8148	5·03 (± 0·1)	Saturn	95·2	0·69
Earth	1·0	5·52	Uranus	14·5	1·60
Moon	0·0123	3·35	Neptune	17·2	2·25
Mars	0·1074	3·92			

Table 2.1 shows that for the terrestrial planets (with the possible
exception of Mercury, for which both the mass and radius are
subject to large uncertainty) the order of increasing mass,
namely moon, Mars, Venus, and Earth, exactly agrees with the
order of increasing density, despite the fact that the solid radius
of Venus is still uncertain to within possibly as much as 50 or
100 km. But turning to the values for the outer planets, not
only does no such ordering with mass emerge but the mean
densities are all of a smaller order than those of the terrestrial
planets, and it is impossible to suppose that they are rocky
bodies even approximately similarly constituted to the terrestrial group. It is not our purpose here to consider their composition in detail, but as has been seen in Chapter 1 the dust-

disk hypothesis of their origin suggests that even these bodies probably have small cores, possibly of terrestrial mass in order of magnitude and similarly composed to the inner group of planets, but that the great bulk of their mass is contributed by the very lightest gases, helium and hydrogen. There appears to be no other way that a massive body such as Saturn could exhibit a mean density considerably less than that of water at ordinary atmospheric pressure.

The pressures deep within the great outer planets, certainly for Jupiter and Saturn, are hundreds of times greater than within the terrestrial planets—for the Earth these are of the order of a million atmospheres—and this must make them extremely complex objects because of the possibility of several phase-changes occurring over the whole range of these extremely high pressures. There is, as we shall see, abundant evidence that even the Earth itself has a complicated internal structure, but there can be little doubt that the terrestrial group of planets together with the moon are structurally simpler than the outer group. From a cosmogonical viewpoint one of the principal questions they present is whether they can be regarded as having identical or even closely similar compositions. For many years, the seemingly high average density of the Earth, as compared with the moon and Mars, for example, was thought to imply a systematic difference of composition. It was suggested that at an earlier stage iron had been able to separate out within some larger primordial system, and that by some means the Earth had secured a greater proportion of this than had such bodies as the moon and Mars. A possibility of this kind seemed supportable on the hypothesis that these bodies originated in a molten form, permitting the separation of denser materials such as iron and nickel from the lighter materials.

Evidence was first discovered early in the present century for the existence within the Earth of a central liquid core extending nearly 55 per cent of the distance from the centre to the surface. The actual radius of this core has been determined as 3473 km, compared with the overall (spherical) radius of the

Earth of 6371 km. Long before the liquid character of the core had been established from seismic considerations, it had been conjectured from the high mean-density of the Earth as a whole that it might contain heavy materials in its central parts, much denser than those of average value near the surface, and iron as an abundant element must have seemed the most likely possibility. Moreover at the time concerned, the existence of the so-called Curie-point—the characteristic temperature above which substances lose any ferromagnetic properties (about 750°C for iron and only about 350°C for nickel)—had not been recognized, nor had the high temperature of the central regions of the Earth, presently believed to be of the order of a few thousand degrees. In addition, meteorites seemed commonly to be composed of iron. But in fact, if only those actually seen to fall are considered, the proportion of ferrous meteorites is probably less than 5 per cent by number; after their fall, stony meteorites are less likely to be recognized as of cosmical origin than are ferrous ones. Also it seemed perfectly natural that if the Earth began as an entirely molten body, the heavy iron should settle to the centre.

Whatever the reasons, there can be little doubt that the notion that the Earth possesses a central iron or iron–nickel core, accounting for almost one-third of its total mass, became a widely accepted conclusion, and with it came the implication that such bodies as the moon and Mars, whose mean densities were reasonably accurately calculable, could not possess any similar amount of iron core. The problem then apparently arose why such differences in composition should subsist. But if this group of inner planets accumulated from a disk of dust, itself captured by the sun from interstellar space, it would require some very exceptional processes to segregate any particular element such as iron into the special zone of the disk where the Earth accumulated. Besides, there is no certainty at all that any iron contained in interstellar dust exists as free iron, and it seems more probable that it would be in the form of compounds. Again, a certain amount of iron also occurs in the remainder of the Earth, and if the core were entirely iron the total quantity within the whole Earth would be at least 35 per cent. This would be far higher than the abundance of iron likely to result from stellar processes leading to the generation of the complete range of elements, and would make the Earth very much of a cosmical curiosity as regards composition. Although this is not an altogether unanswerable argument, the conclusion is sufficiently exceptional to be treated with extreme caution.

Another difficulty with this theory of the composition arises from a seemingly remote question concerning the surface features of the Earth, in particular with the formation of mountains. On the hypothesis of an initially molten Earth, all the subsequent evolution is supposed to be determined by the consequences of cooling, and contraction resulting from this is believed to be capable of producing corrugations at the surface, corrugations that the theory would claim are to be associated with the presence of mountains, especially of the folded and thrust types that form the great majority of orogenic features. This is the 'thermal-contraction theory' of mountain-building. The actual mechanism proposed is not just one of direct contraction, but a kind of differential contraction. If first one imagines an entirely molten Earth, then cooling at the surface would produce a thin solid crust there while the interior would remain liquid. As further cooling solidified the outer part of this, contraction would occur, and the original outermost crust would be too large to fit over the newly solidified part. In practice the process would not take place in separate stages like this. Cooling and solidification would occur continuously, but the effect would be much the same and would lead gradually to a solid exterior too large to fit over the solidifying adjacent lower layers. The outer surface would thus undergo stresses that would eventually exceed the strengths of the materials, and buckling and folding would result.

The associated changes can be thought of in terms of a decrease in the radius of the Earth. If a solid planet comparable with the Earth in mass and size should for some reason undergo radial contraction, then it will be only at and near the surface that any distortion of shape will take place and produce departure from sphericity. The reason for this is that throughout most of the planet the internal pressures far exceed the strengths of materials. Substances such as rocks and iron have strengths of the order of 10^9 dyn cm^{-2}, and the pressure within the Earth rises to this value at a depth of only about 4 km. At about 35 km depth the pressure exceeds 10^{10} dyn cm^{-2}, and throughout by far the greater part of the Earth it exceeds 10^{11} dyn cm^{-2}. Thus in the deep interior, and indeed in all but the outermost 10 or 20 km of the Earth, the pressure dominates the situation, and the strength of the materials, although these remain solid down to a depth of about 2900 km, contributes little or nothing to the support against gravity. But this is not so near the surface; and if for some reason the Earth contracts, the surface layers will buckle and fold just as soon as the extent of the contraction is such as to give rise to forces exceeding the elastic limits of the materials concerned. If for other reasons the surface were already fractured, one layer might slide over another as the easiest way to meet the requirements of overall contraction, and in this way form thrusted mountain ranges. Of necessity, the folding and thrusting would automatically occur at the weakest parts of the surface.

This contraction theory has been extensively investigated by mathematical means, but the numerical values associated with the physical quantities involved are not known with much certainty except possibly for the very surface layers of the Earth, so there must be some element of doubt over the precise amount of contraction that might be expected from this cause. On the other hand, even when all the factors concerned are given what seem to be their optimal values, from the point of view of maximizing the contraction, the decrease in the radius of a molten Earth during its whole development to the present state, supposing that to have been the course of evolution, can only have been about 10 km. The accompanying circumferential contraction, which gives a better indication of the amount of folding and thrusting, would be just over 60 km. But a value as small as this has never been felt to be adequate by many geologists who believe mountains to be due to some process of worldwide contraction.

It is one of the difficulties of theoretical work that there can always be found different schools of thought that maintain contrary conclusions based, so they all feel, on precisely the same range of evidence. At the present time, it may be noted, there are geologists that believe in an expanding Earth, there are some that favour an unchanging Earth, and yet others that favour a contracting Earth. It is in the assessment of the importance and meaning of the evidence that such differences can arise, and as we shall see it could even be that while the Earth in the long run is contracting, periods of intense contraction might be separated by periods of comparatively gentle expansion. However, as the conclusion that will be reached in this essay is of contraction of the Earth on a scale far greater than is capable of being explained by cooling, we take the liberty of quoting what Professors Leet and Judson say on this subject in their great work on physical geology:*

'One generalization on which all geologists agree is that mountain-building involves a reduction in the surface area of the globe—a shortening of the distance between points on the surface—and that all mountain-building is the product of a single mechanism—*squeezing by horizontal compression*. But when it comes to what *causes* the squeezing, there is no general agreement.'

Then again, the late Dr. G. M. Lees, who spent his life in the field as a working geologist and spoke with the highest authority on the practical side of such questions, was inflexible in his conclusion that one time the Earth must have been far larger than at present, possibly by more than 1000 km.†

Evidently then, there have been voices raised against the thermal contraction hypothesis even in its own right, and quite apart from whether the Earth as a whole was ever molten.

* *Physical Geology*, p. 358, Prentice Hall (1961).
† The evolution of a shrinking Earth. *Q. Jl geol. Soc. Lond.* **109**, 217 (1953).

Accordingly some attempts have been made to look for other causes of mountain-building. Even the ancient idea of the separation of the continents from each other by lateral movement across the surface of the globe—continental drift—has recently been revived and amplified by the so-called 'convection theory' of the wholesale movement of material within the Earth. At first sight one might suppose that a convection theory would concentrate on the liquid core for its realm of action, but here this is not so, and the convection in fact is assumed to occur in the solid mantle. To sketch the theory briefly, it involves two main requirements. First, because of postulated slight differential effects of asymmetry of density or temperature (or both) suitably located within the mantle, extremely slowly moving currents, with speeds of the order of 1 cm per year or less, are set up and take the form of well-arranged convection-cells occupying the volume bounded internally by the liquid core and externally by the surface of the Earth. Secondly, the theory requires a growing liquid core, and the reason proposed for this growth is that iron is actually draining out of the solid mantle into the core, because the iron is denser, and, having a lower melting-point than rock, liquifies on joining the core or even before this.

The theory then maintains that in the early stages of development, when the liquid core would be small, a certain number of circulatory convection cells would automatically become set up, more or less filling the volume of the solid mantle, the flow in them being upwards at some regions and downwards at others. But with the growth in size of the core, a stage would come when this arrangement and number of cells would no longer represent a configuration that could persist steadily. Instability would then set in, and eventually a new arrangement involving a larger number of cells would come to subsist. The period of adjustment from a given number of cells to a higher number is regarded as involving serious upheaval of the interior with slighter manifestations at the surface in the form of mountains. The systematic drag of the currents on material near the surface might operate to produce mountains at all times, and possibly might even cause the continents themselves to drift.

Ingenious as the theory is, so far it has been put forward mainly in descriptive terms, and much remains to be done towards demonstrating mathematically that the proposed effects would even be of the order of magnitude required. There is difficulty in establishing whether even such small currents would occur in solid material, for there seems no way of deciding whether small forces applied for very great lengths of time will cause 'solids' to flow. Recent evidence from the detailed motion of artificial satellites suggests that the Earth is slightly more spheroidal in form than is consistent with its present rotation rate. As it would require considerable strength to maintain this non-equilibrium shape, which incidentally could be interpreted as a relic of a stage of faster rotation, this interpretation would be evidence against the notion of convection within the solid mantle and outer shell.

On the other hand, it is possible to argue that the departure from equilibrium may be due to the convection currents themselves, though it seems highly questionable whether such minute currents would in fact lead to the production of the Himalayas and the Alps, and the huge ranges of the Rockies and Andes, and other great mountain ranges of the world. Are such features really what the surface response to minute circulation currents is going to be? At all events, it has yet to be shown that the response would be as impressive as are the mountains.

There is also the question, impossible to check directly, whether solid masses of iron are distributed in the mantle, and, if there are, whether they could in fact make their way down through it, as if it were a highly viscous liquid, to the core. This again is a matter of whether there is a threshold of strength that has to be exceeded before one solid will move through another of different density under the force of local gravity, or whether long-sustained forces, however small, will prevail to produce motion. It is known that materials, such as pitch and certain kinds of glass, which are apparently solid, do in fact flow very slowly, but whether rock and iron yield in the same way to long-sustained small forces is still an open question. If the Earth accumulated from meteorites, which is an idea proposed by one school of thought, then there might well be sizable chunks of free iron, though scarcely to the extent of 30 or 40 per cent in total mass. But much detailed evidence seems to suggest that the meteorites do not represent original pre-planetary material but are a later product of the solar system, and that they have at some stage been at considerable pressure and temperature within some far larger mass such as a planet. The meteorites themselves remain one of the most perplexing and inscrutable problems of the solar system, and it would probably be a great mistake to dispose of the mystery of the source, or sources even, of meteorites by regarding them simply as a negligible residue of original solar-system material.

As far as the internal structures of the Earth and other terrestrial planets are concerned, the convection theory strives for a higher degree of refinement than seems attainable at present, for even the first-order theory of a static Earth is scarcely more than beginning to be understood. The situation is somewhat analogous to that existing half-a-century or so ago in regard to stellar structure. At that time Eddington attempted a theoretical study of the famous variable stars called Cepheids, which appear to be undergoing not only observed rhythmical changes of luminosity but accompanying pulsations in their sizes, but he soon came to the conclusion that it was almost useless trying to grapple with a problem of this order unless it was first understood how an ordinary static star, presumably of much simpler structure, could maintain its steady luminosity and size. Several decades were to pass before even this less ambitious question could be regarded as satisfactorily answered. So in the present case it may be that the convection theory is premature, and that the first necessity is to understand much more fully the theory of the structure of static perfectly symmetrical planets.

Besides all these considerations, the convection theory appears to have ignored the question of what happens to the size of the Earth if iron, which is more compressible than rock, is transferred downwards always to regions of higher pressure. Might not this alone cause an Earth so constructed and so developing to contract? But if the existence of masses of free iron is questioned, and the hypothesis of the formation of planets from dust would certainly raise grave doubts, there would nevertheless arise the problem of the early structure of the Earth as an all-solid body. What can be done about investigating this? Again the problem is closely analogous to that of stellar structure, where the central problem may be stated: Suppose one is given a definite mass of material of stellar order, and suitable information as to the composition (this latter, by the way, does not necessarily mean the *name* of the material, such as iron or silicon, but the numbers representing its relevant properties within some equation-of-state for a star; that is, the atomic weights of its particles and the physical laws that the material obeys); then how luminous would such a star be, and how large would it be? In brief, given the mass and composition, what would be the luminosity and radius? The fundamental part of the theory of stellar structure concentrates on these questions, and it is found that in order to answer them satisfactorily the whole internal structure of a star has to be taken into consideration.

Where the original all-solid Earth, or a postulated all-solid Earth, is concerned no important intrinsic luminosity would be expected, and the central initial problem is therefore to find how large it would be. Where the mass is concerned, if it is supposed that the process of formation is complete, then the value cannot be significantly different from the present mass, namely $5 \cdot 976 \times 10^{27}$ g. But the composition presents a more difficult problem. If some uniform density throughout were assumed, calculation of the radius would be a trivial matter, but in fact whatever the material it will have elastic properties at the pressures produced by the overall gravitational field of the planet. Clearly it will be compressed most at the centre, where the pressure is greatest, and not at all at the surface, where the pressure is zero or at least negligible. What are really required to be known in order to calculate the radius are therefore the uncompressed densities and the elastic properties. Thus information about the original substance means, for the present purpose, numerical information of the foregoing kind, together with some relation expressing how the density changes with applied pressure.

The properties of the deep interior of the Earth would indeed be difficult to determine were it not for the circumstance that earthquakes occur frequently in various parts of the Earth to varying degrees. When a large earthquake takes place it can send waves reverberating through the entire planet, and the arrival of these at distant stations distributed over the globe can be recorded and accurately timed. There are several types of earthquake-waves, of which the two principal ones are the so-called primary, or P-waves; and the secondary, or S-waves. When energy is propagated by means of P-waves, the oscillation is practically in the direction of travel of the rays. With this picture in mind they are sometimes called push-waves, and they are of such a nature that they can be transmitted in both solid and liquid material. But for the S-waves, sometimes called shake-waves, the oscillatory disturbance at any point is transverse to the direction of propagation. Waves of this type can occur only in a solid, because they depend on the rigidity of the material for their transfer from one point to the next. They cannot occur in a liquid, and it is precisely from the absence of S-waves in the core of the Earth that its liquid character is inferred. The two types travel at different speeds in a solid, and the velocity of the primary waves v_P is always essentially faster than that of the secondary waves v_S. The theory of elasticity relates these speeds to the physical properties of the material, in particular to the density ρ, the rigidity μ, and the bulk-modulus k, which last quantity turns out to be of special importance to the problem of finding the radius of an all-solid Earth. The relations for these speeds are

$$v_\mathrm{P}^2 = (\tfrac{4}{3}\mu + k)/\rho, \qquad v_\mathrm{S}^2 = \mu/\rho, \tag{2.1}$$

so that the bulk-modulus k is given by

$$k/\rho = v_\mathrm{P}^2 - \tfrac{4}{3} v_\mathrm{S}^2. \tag{2.2}$$

But it is also the case, for chemically homogeneous material, that

$$k = \rho \frac{dp}{d\rho}. \tag{2.3}$$

These relations, together with the equation of hydrostatic equilibrium, would enable the internal structure to be determined and the values of k, p, and ρ to be found at all depths in the Earth provided that the values of v_P and v_S can be found; and it is precisely this information that the seismic data of travel-times of earthquake waves eventually supply. Over the past three-quarters of a century, a gradually improving detailed picture of the interior of the Earth has been built up, and the pressure, density, and bulk-modulus, together with other elastic properties of the material, have been determined. The resulting information is such that the Earth can be thought of as consisting of seven regions of differing properties. These have been labelled A, B, C, D, E, F, and G for reference purposes by Bullen: starting at the outer surface they are as follows:

Region A: consisting of the *crustal layers* and extending from the outer surface to a depth of 33 km.

Region B: forming (with A) the *outer shell* extending down to 413 km where the so-called 20°-discontinuity first occurs. (The pressure at this level is 0.141×10^{12} dyn cm^{-2}.)

Region C: a transition region representing the outer part of the mantle down to about 984 km.

Region D: the main part of the mantle extending down to depth 2898 km, where a sharp change to the liquid state occurs. (The pressure at this core–mantle boundary is 1.37×10^{12} dyn cm^{-2}.)

Region E: the outer liquid core extending to depth 4982 km.

Region F: a shallow transition region in liquid form from 4982 km to 5121 km.

Region G: the innermost core of radius 1250 km (which may be in solid form).

As the depth through all these successive regions increases so also does the actual density of the material, but not always continuously. The plot of density against depth is shown in Fig. 2.1. The density of the crustal layers is only about 2·7 g

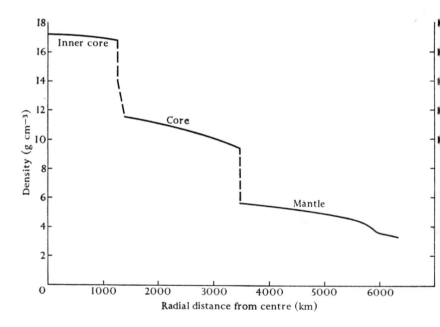

Fig. 2.1. Variation in density with radial distance within the Earth.

cm⁻³, but at the top of region *B* it is about 3·3 g cm⁻³; and from there on down through *B* it increases steadily to a value 3·6 g cm⁻³ at 413 km depth, the 20°-discontinuity. The rate of increase itself now suddenly becomes greater for a time through C, and then is slower through the mantle D, at the bottom of which the density is $5 \cdot 7$ g cm^{-3}. But now a sudden large increase occurs, by some 66 per cent, so that at the outer part of the liquid core the value is about $9 \cdot 4$ g cm^{-3}, and below that it continues to increase. The density at the centre of the innermost core may be as high as 17 (± 5) g cm^{-3}, but the actual value remains uncertain, as also does the nature of this core-within-the-core. It is possible that it represents the heavier substances that have settled out of the liquid core as a whole. If the core is no more than a denser phase of mantle material, it is possible that the innermost core consists of heavy metals drained to the lowest level. That the mass of the innermost core is just about 7 per cent of the mass of the whole core would be more consistent with this possibility than the notion that the whole core, with over 30 per cent of the mass of the Earth, consists of heavy metals such as iron and nickel.

In calculating travel-times, it is necessary to take account of the differing properties of all these seven regions; indeed, it is by the reverse process that their existence has been ascertained. For calculating the radius of an initially all-solid Earth, or the radius of an Earth with a liquid central core of mass different from the present value ($1 \cdot 876 \times 10^{27}$ g), it is possible, however, to regard the Earth as consisting of at most three main zones; and for an all-solid Earth just two main zones. To see that such a simplification is permissible, it may be noted that although the innermost core has nearly one-fifth the radius of the Earth, it contains only just over 2 per cent of the mass and occupies less than $0 \cdot 75$ per cent of the entire volume, whereas it is found that much greater percentage volume changes result for an all-solid Earth as compared with the present one. But for inter-mediate models with some liquid core, it is possible to use the averaged properties of the entire core: regions E, F, and G: determined in such a way that a homogeneous core of these average properties and of the present mass would have the same radius and pressure at its boundary (with the mantle) as exist in the actual Earth. By such means a more accurate result will emerge than if no allowance for a region corresponding to G were made in the calculations.

Then again, the volume of region A is only about 1·5 per cent of that of the whole Earth. If it is duly averaged with the much larger region B, the two together can be regarded as constituting a single outer-shell zone with only minute resulting inaccuracy. The same can be done for regions C and D, whose average physical properties can be adopted as representative of a single mantle zone. In fact, when all the various physical properties are plotted against depth, or against pressure, the resulting curves all strike the eye as dividing into three main parts, corresponding to $A + B$, $C + D$, $E + F + G$, and themselves suggest this simplification into three main zones.

For the present purpose, the most interesting of all these curves is that of the bulk-modulus k against pressure p, which is shown in Fig. 2.2. As equation (2.3) shows, the quantity k is itself of the dimensions of pressure, and it is seen from Fig. 2.2 (p. 61) that in each of these three main zones k is with high accuracy a linear function of p; thus

$$k = a + bp, \tag{2.4}$$

where a and b are certain constants. It is convenient to adopt the unit of 10^{12} dyn cm^{-2} for k, a, and p, since these quantities take values of this order throughout almost the entire Earth. The graphs of k against p in the core and mantle are as near parallel as the eye can distinguish, which means that the value of b is practically the same in these zones; it is in fact very close to 3·5. The values of a are substantially different, however, and the fact that an extension of the straight line corresponding to the core would come below that of the mantle means that the core-material is more compressible than the mantle-material for a given pressure.

Determination of the values of a and b appropriate to each zone can be made by calculating the straight line of closest fit to the available (k, p)-values. A relation of the form (2.4) when

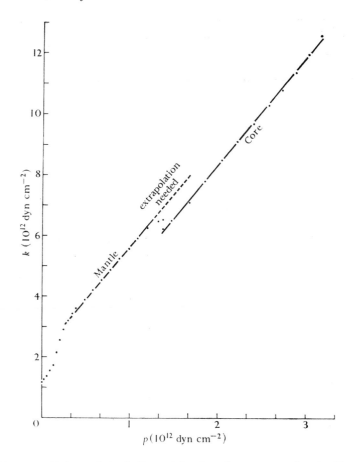

Fɪɢ. 2.2. Values of the bulk-modulus k and pressure p for the Earth. The lines are $k = a + 3 \cdot 5p$ for the mantle and core.

found in this way is purely empirical, and the observed values of (k, p) could equally well be represented by a continuous smooth curve drawn through all the (k, p)-points. Nevertheless, there is the possibility that a linear relation with $b = 3 \cdot 5$ represents an actual physical law, though one not yet established. With this in mind, it would also be possible to set $b = 3 \cdot 5$ from the outset, and then calculate the single quantity a to give the best fit in each zone. Both these calculations have been made, and the results are set out in Table 2.2.

Table 2.2

	a (10^{12} dyn cm^{-2})	b	Range of k (10^{12})	Mean-square error of k
Outer shell (0–413 km)	1·109 1·167	4·349 3·5	1·16–1·71	0·009 0·040
Mantle (413–2898 km)	2·222 2·151	3·390 3·5	3·19–5·90	0·024 0·037
Liquid core (2898–4982 km)	1·286 1·342	3·524 3·5	6·20–12·6	0·066 0·068

In considering Table 2.2, it has to be remembered that the values of k and p on which it is based are themselves not perfectly accurately known, so that the closeness of the values of b to 3·5 in both the mantle and liquid portion of the core, especially in the latter, is certainly remarkable. Also, it is seen that the accuracy of representation of the data, shown in the last column, when b is taken at exactly 3·5 is not much affected, and the changes even in the mantle are probably less than the intrinsic inaccuracy of the k-values. In the outer shell, b differs more widely from 3·5, but when this precise value is adopted the resulting errors in the linear representation of k, though increased, become only of much the same order as those occurring in the other zones. But what is important here is that whichever of these two slightly different representations of the data is used in calculating the dimensions of any model planet, the two sets of results are found to differ only by extremely small amounts, far less indeed than the accuracy with which most of the quantities concerned may ever be known. This strongly suggests that if in the calculations values given by a continuous curve representing the individual (k, p)-points accurately were used, results differing only by similar minute amounts would emerge. Mathematically, for the purpose of getting simple analytical relations revealing how the properties depend on the total mass M and upon a, it is convenient to take $b = 3\cdot5$, but where machine calculations are concerned either set of values of a and b can of course be coped with equally readily.

Before proceeding to describe such calculations, an extremely interesting further result emerges from the numerical data for k, p, and ρ when equations (2.3) and (2.4) are taken together. In each of the three homogeneous zones, this pair provides a differential relation

$$\rho \frac{dp}{d\rho} = a + bp, \qquad (2.5)$$

with known but different values of a (and possibly of b if the precise value $3\cdot5$ is not adopted) in each zone. This equation integrates quite simply to the form

$$p = \frac{a}{b}\left\{\left(\frac{\rho}{\rho_u}\right)^b - 1\right\}, \qquad (2.6)$$

where ρ_u is a constant of integration, which in fact corresponds to the uncompressed density of the material in the zone concerned. Clearly, this equation can now be solved for ρ_u for every set of values for p and ρ available in each zone, and all the resulting values should be practically the same. Table 2.3 shows the uncompressed densities that actually emerge in this way with $b = 3\cdot5$.

Table 2.3

Outer shell	ρ_u	$3\cdot298 \pm 0\cdot006$ g cm^{-3}
Mantle	ρ_u	$3\cdot975 \pm 0\cdot015$ g cm^{-3}
Liquid core	ρ_u	$6\cdot107 \pm 0\cdot004$ g cm^{-3}

The outer-shell value of $3\cdot298$ g cm^{-3} is, of course, not the actual surface-density found for continental-type material, for example; but neglecting all the surface peculiarities the figure is what the value would ideally be. Then in the mantle there is a jump to just under 4 g cm^{-3}. This has long been regarded as the result of a phase-change to a denser crystal form produced by high pressure. That a is much larger in the mantle than in either the outer shell or the core means that the mantle-type material is much more incompressible than either of the other two types.

But the really interesting thing here is the value that emerges for the uncompressed density of the liquid core of only just over 6 g cm^{-3}. This is so much less than the value for iron that it must raise serious questions whether in fact the liquid core can be regarded as composed of such heavy metals at all. When the calculations here described were first attempted, it was on the basis that the growth of the core resulted from iron draining down through the mantle, as explained earlier, and it came as a surprise that the uncompressed density should be so low. Nor can the difficulty be got round by supposing the liquid core to be some high-pressure phase-modification of iron, for then the apparent uncompressed density would obviously be higher than that of any low-pressure phase of iron. The obvious alternative lies in a development of the hypothesis proposed by Ramsey nearly twenty years ago to the effect that the core represents a phase-change of mantle material produced by high pressure. But in this form, with dependence solely on pressure, the hypothesis clearly would not lead to a dynamic evolving Earth, and it seems more probable, as will be discussed later, that the phase-change is the combined result of both pressure *and temperature*, with the latter rising steadily in the deep interior as a result of the release of energy by radioactive materials. A sufficiently high temperature will liquefy any substance at ordinary low pressure, and will indeed eventually ionize it, but it is a fairly novel suggestion that a sufficiently high pressure would do the same, the pressure required depending on the temperature. As yet, static pressures of the order of those prevailing in the core have not been attained in the laboratory, but even should they be achieved it will be necessary also to impose temperature conditions corresponding to those in the deep interior—possibly a few thousand degrees—if the experiments are to be applicable to the actual situation and test the present hypothesis. On general grounds, it seems certain that matter does not have infinite strength, and hence that if it is subjected to sufficient pressure the collapse of successive electron-shells would result. Thus the core material could well be a liquid ionized form of mantle material, and then would be of metallic form equally capable of meeting any requirements for the production of a magnetic field as it would be if it were molten iron.

Let us return now to our original objective, which was to calculate the dimensions of an initially all-solid Earth. It is here that the value of a relation such as $k = a + bp$ or its integrated form (equation 2.6) is especially important, for the fact that k depends only on pressure means that the relation can be applied with some confidence to earlier states of the Earth. The release of radioactively produced energy means that the internal temperatures of the planets will have risen from their initial values; indeed, it now appears probable that evolution of the Earth and similar planets arises from this source of energy. But the fact that k is practically independent of temperature makes it possible to apply equation (2.6) not only to earlier stages of the Earth but also to other planets, on the hypothesis that they are of similar composition. The reason for absence of temperature-dependence in k appears to be that the temperatures concerned, which are of the order of 10^3 °K, would produce volume changes comparable with those produced by pressures of the order of only 10^9 dyn cm^{-2}, whereas throughout almost the entire Earth the pressures are of the order of 10^{12} dyn cm^{-2}, and so the dependence is purely on pressure. The same holds for the other terrestrial planets; indeed, even in the moon, apart from the outermost few tens of kilometres, the pressures much exceed the strengths of the materials.

This also means that the support of the material against internal gravity is provided almost entirely by the pressure, and the equation of hydrostatic equilibrium therefore holds with high accuracy. This equation may be written

$$\frac{dp}{dr} = -g(r)\,\rho(r) = -\frac{G\,m(r)\,\rho(r)}{r^2}, \qquad (2.7)$$

where $g(r)$ is the acceleration of gravity, $\rho(r)$ the density, at distance r from the centre of the planet, and $m(r)$ the total mass within distance r. Equations (2.6) and (2.7), together with the relation

$$dm(r) = 4\pi r^2 \rho \, dr, \qquad (2.8)$$

are sufficient to determine the three quantities p, ρ, and m at all points interior to a planet of given mass provided that the appropriate values of a and b are adopted for its several zones. The equations are non-linear and the solutions have to be found by machine integrations, but there is no point of difficulty in determining the solutions appropriate to any assumed conditions.

For an all-solid Earth, it turns out that the central pressure is only about 20 per cent greater than the value at the base of the mantle now, so it would be expected that at this stage the planet would consist only of two main zones, namely an outer-shell zone and an interior mantle zone. To effect the calculations, the amount of mass in the form of mantle material can be regarded as a parameter m_c, and for any selected m_c a model configuration for an all-solid Earth is arrived at uniquely. Table 2.4 shows values of the principal properties for a range of values of m_c.

Table 2.4. *All-solid models of the Earth*

m_c (10^{27} g)	ρ_0 (g cm^{-3})	p_0 (10^{12} cgs units)	r_c (km)	p_{int} (10^{12} cgs units)	r_s (km)	ρ_m (g cm^{-3})
5·0	5·756	1·630	6251	0·1578	6775	4·587
5·1	5·756	1·631	6299	0·1412	6769	4·601
5·2	5·757	1·632	6346	0·1247	6762	4·615
5·3	5·758	1·633	6392	0·1083	6754	4·630
5·4	5·758	1·633	6439	0·0920	6747	4·645
5·5	5·758	1·634	6485	0·0758	6740	4·661
5·6	5·759	1·634	6530	0·0597	6732	4·677
5·976	5·759	1·635	6700	0·0	6700	4·745

m_0 = mass of central region consisting of mantle material
ρ_0 = density at the centre
p_0 = pressure at the centre
r_c = radius of the central region
p_{int} = pressure at interface where outer shell starts
r_s = radius to the outer surface
ρ_m = mean density of the whole model

It is seen from Table 2.4 that over the range taken for m_c, which is from $5 \cdot 0 \times 10^{27}$ g to $5 \cdot 976 \times 10^{27}$ g (the total Earth mass), the central pressure and density change very little, and the mean density only slightly more so. An initially all-solid Earth would of course have taken up a unique configuration, and the parameter that settles which particular model is appropriate is in fact the interface pressure, p_{int}, shown in column 5, which varies considerably over the range tabulated. At the present time in the Earth, this pressure has the value $0 \cdot 141 \times 10^{12}$ dyn cm^{-2}, but it is known that the phase-change here involved is highly sensitive to temperature, and the greater the temperature the greater the pressure required to effect it. The present internal temperature of the Earth is likely to be much higher than it was originally because of radioactive energy release, and this makes it reasonably certain that only values of m_c in excess of about $5 \cdot 1 \times 10^{27}$ g, for which $p_{int} = 0 \cdot 1412$ dyn cm^{-2}, are likely to be relevant. But it is seen from the next column that the overall radius r_s for every possible such model substantially exceeds the present Earth-radius of 6371 km, and even for the smallest all-solid model consisting entirely of mantle material the excess is as large as 329 km.

Thus almost whatever the initial value of p_{int} may have been, equal to or less than the present value, in an initially all-solid planet, the calculations show that such an Earth would have been of radius at least 350 km greater than the present value. This is more than a whole order of magnitude greater than would result from thermal contraction, and is possibly sufficiently great to satisfy geological requirements at long last. In fact, it is the change of *surface area*, not of circumference, that would have to be tucked away by folding and thrusting, and since for a sphere of radius R this quantity is $\delta(4\pi R^2) = 8\pi R \, \delta R$, the resulting change when so measured is far more impressive. Indeed, it implies a decrease of surface area of about 60 *million* square kilometres, more than a tenth the present area of the globe, whereas the 10 km decrease for the thermal contraction hypothesis can lead to a mere $1 \cdot 6$ million square kilometres. The implied circumferential shortening would be about 2500 km on the present basis. It is scarcely necessary to stress the possible importance of these results to the problem of formation of mountains at the surface. In Table 2.5 are shown successively the interface pressure, the change of radius as compared with the present Earth, the corresponding decrease of superficial area, and the depth of the outer shell for the series of models of Table 2.4 for $m_c = 5.1$ to 5.6 ($\times 10^{27}$ g).

Table 2.5   Decrease of radius and surface area; and depth of outer shell

m_c $(10^{27}$ g)	p_{int} $(10^{12}$ dyn cm$^{-2})$	$r_s - r_E$ (km)	Decrease of surface area $(10^6$ km$^2)$	Depth of outer shell $(r_s - r_c)$ (km)
5·1	0·141	397	65·5	470
5·2	0·125	390	64·4	416
5·3	0·108	383	63·1	362
5·4	0·092	376	62·0	309
5·5	0·076	368	60·7	255
5·6	0·060	360	59·3	201

From the circumstance that as the internal temperature increases the transition from outer-shell to mantle form of the material requires higher pressure, it would follow that radioactive heating would cause the outermost mantle material to revert back to outer-shell form, and the interface therefore to move inwards with an accompanying *increase* of overall radius. If, for example, the interface pressure were to have increased from 0·06 to 0·10 ($\times 10^{12}$ dyn cm^{-2}) before any melting of the central regions of the Earth commenced, the overall radius would have increased from about 6730 km to about 6750 km, while the depth of the outer shell would have increased from about 200 km to 335 km. This expansion by about 20 km would be accompanied by an *increase* in surface area of about 3·4 million square kilometres and of circumference by about 126 km. But this initial increase is less than one-tenth of the subsequent contraction (to the present value) associated with the growth of the liquid core. A period as long as 10^9 years or more may have elapsed before the central temperature became sufficiently high for liquefaction to occur, and any expansion during this stage may well have operated to crack and sunder the surface layers to a depth of at least a few kilometres, and thereby produce conditions favourable to the thrusting of adjacent layers over each other during subsequent large-scale contraction.

One final point before leaving this part of the problem is that the central pressure for all these models listed in Table 2.4 (p. 66) is about $1·63 \times 10^{12}$ dyn cm^{-2}, and this is less than 20 per cent greater than the present value at the base of the solid mantle. Thus the calculations involve extrapolation by only this moderate amount of the linear law for k for the mantle-type material. For the actual liquid core to represent a denser phase of mantle material, and for this now to occur at a pressure of only $1·37 \times 10^{12}$ dyn cm^{-2}, is also explicable by a lower initial temperature than at present obtains at the core-mantle boundary. This is consistent with the hypothesis maintained here that the phase-change results from the combination of both pressure and temperature: the higher the temperature, the lower being the pressure required to produce the transition, and vice versa.

The amount of contraction implied to reach the present size may seem large, but if distributed evenly over several aeons it would scarcely be noticeable at all. A continuous steady contraction of 350 km spread over $3·5 \times 10^9$ years, to take round figures, amounts only to 0·01 cm per year—one-tenth of a millimetre annual decrease of radius! It is not likely, however, that the change has occurred continuously at this rate, for it seems more probable that, as the interior of the Earth contracted, stresses would gradually be built up in the solid surface layers, and these would have no serious effect until they attained to values somewhere exceeding the strength of the materials. At this stage, a rapid period of adjustment would come about, and would gradually die down only as the various stresses were relieved. Thereafter, further internal contraction would take place, again gradually building up fresh stresses in the surface layers.

Several epochs of mountain-building have been traced right

back in time by geologists. There have been at least three major periods well established even for post-Cambrian times, which means as recently as within the last 500 million years, and numerous earlier ones have occurred ranging through some thousands of millions of years, with their times of greatest activity, it seems, separated by intervals of the order of a few hundred million years. Nor is there any reason to suppose that the much slighter expansion resulting from increasing depth of the discontinuity between mantle and outer shell would cease, and it could well be that between periods of most active mountain-building some slight expansion of the Earth would occur. Thus there might well be indications of both expansion and contraction at the outer surface, but the latter would necessarily dominate in the long run and by a large factor.

Let us consider next how this original all-solid Earth would evolve. The graph of temperature against radial distance within the Earth under radioactive heating is at first sight rather curious. Suppose for simplicity that the initial temperature was the same everywhere, and equal to the temperature at the surface maintained by the sun, so that the graph of T against distance r is a straight line. Then after a given interval of, say, 5×10^8 years, the central temperature will have risen, but the form of $T(r)$ remains practically a horizontal straight line until a distance from the surface is reached small compared with the radius; the curve then turns down sharply and runs to the fixed surface value. After 10^9 years, say, the central temperature would be greater still, but $T(r)$ remains almost horizontal, though with the turn-down point slightly deeper in; and so on. The situation is shown schematically in Fig. 2.3. If in this figure $T = T_m$ is the initial melting-point temperature at the centre, then the Earth would remain all-solid until radioactive energy had brought the central temperature up to this value. It is not possible at present to say exactly how long this would take, because not only are the thermal properties of the material at these extremely high pressures uncertain, but the quantities of radioactive materials that can be assumed to be in the Earth are

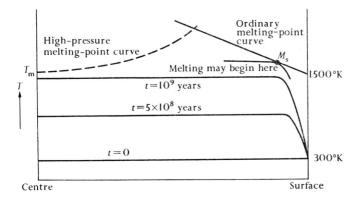

FIG. 2.3. Variation of temperature with radial distance within the Earth under radioactive heating. The three curves illustrate conditions initially, after 5×10^8 years, and after 10^9 years.
T_m, initial melting-point temperature at the centre; M_s, point of intersection of temperature curve and melting-point curve.

also subject to uncertainty. However, the requisite temperature is likely to have been no more than a few thousand degrees, and it is possible that an interval of order 10^9 years elapsed before this stage was reached.

The change of melting-point temperature T_m with pressure is determined by the so-called Clausius–Clapeyron equation, which by means of (2.7) (p. 65) can be written

$$\frac{dT_m}{dr} = -\frac{T_m}{L}\left(\frac{\rho_S}{\rho_L} - 1\right)g(r), \qquad (2.9)$$

where L is the energy per unit mass required for the phase-transition from solid to liquid, and ρ_S and ρ_L are the densities of the solid and liquid phases. For the moderate pressures that occur near the surface the solid form has greater density than the liquid form, so that $\rho_S > \rho_L$, and the right-hand side of (2.9) is essentially negative, which means that, with increasing depth below the surface (that is for $dr < 0$), the melting-point *increases*, as indicated in Fig. 2.3. Thus, as the internal temperature rises, it appears that the shoulder on the curve would first

6—M.S.S.

intersect the melting-point curve at some point such as M_s within perhaps a few hundred kilometres of the outer surface.

On the other hand, if the liquid core is a phase-change of mantle material, then the values of Table 2.3 show that it would now be the liquid form that had the greater density, and the sign of the right-hand side of (2.9) would be positive in regions where the conditions of pressure and temperature are such as to produce this kind of phase-change. Thus the melting-point would *increase* outwards now from the centre, and the melting-point curve here would be of the form shown by the broken line in Fig. 2.3, just the opposite of the situation for low pressures. If it were the case that $\rho_L < \rho_S$ throughout the Earth, the melting-point would continue to increase all the way from the surface to the centre, and there would be difficulty in explaining how the central regions could come to have been melted at all, since the centre would be the very last place that a point such as M_s (Fig. 2.3) could occur. But on the phase-change hypothesis of the nature of the liquid core it is plain that, for the region of the Earth to which this applies, melting would first occur at the centre. Until more is known of such processes, it is not possible to say whether or not melting might have already occurred in a layer at some such depth as the point M_s sufficiently close to the surface for the phase in which $\rho_S > \rho_L$ to hold. But this possibility does not in fact affect the general character of evolution that will take place once the melting has commenced at the centre.

When it comes to computing static models of an Earth with liquid core of arbitrary mass, m_c say, we are faced with the possibility of three-zone models: core, solid mantle, and outer shell. If the interface-pressure between mantle and outer shell at earlier stages were different from the present value, which as has been seen would probably be the case, it would be necessary to know this pressure for any given m_c; otherwise we would arrive at a doubly-infinite series of models corresponding to the two parameters m_c and p_{int}. But the actual Earth can of course have evolved only along a single linear series, corresponding to increasing m_c, and at present it becomes necessary to make some empirical assumption for the value of p_{int} corresponding to any given m_c. To meet this requirement, the following simple linear relation was adopted:

$$p(\text{interface}) = 0 \cdot 08003 + 0 \cdot 0325 \, m_c, \qquad (2.10)$$

wherein the unit for p_{int} is 10^{12} dyn cm^{-2} and that for m_c is 10^{27} g. For the present value of the core-mass, which is $1 \cdot 876 \times 10^{27}$ g, this agrees with the existing interface-pressure of $0 \cdot 141 \times 10^{12}$ dyn cm^{-2}, while for the original all-solid Earth at the time of liquefaction it gives a value of approximately $0 \cdot 08 \times 10^{12}$ dyn cm^{-2}, qualitatively consistent with the view that at lower temperatures this phase-change can be effected by lower pressures. The values of a and b, in the relation between pressure and density, now take successively the three sets of values given by Table 2.2 (p. 62). The uncompressed density will take the value $6 \cdot 26$ g cm^{-3} in the liquid core, thereby making some allowance for the possibility of an innermost core, as explained earlier, and will be equal to $3 \cdot 975$ g cm^{-3} in the mantle, thus bringing the radius of the present Earth model, with $m_c = 1 \cdot 876 \times 10^{27}$ g, into agreement with the actual value.

The results of the calculations are shown in Table 2.6, in which the values of the following quantities, in addition to those tabulated in Table 2.4 for two-zone models, are given:

p_{12} = pressure at core–mantle boundary,
r_m = radius to outermost part of mantle,
p_{23} = pressure at boundary between mantle and outer shell,
m_{12} = combined mass of core and mantle,
f = moment of inertia factor (so that $I = fMr_s^2$),
fr_s^2 = moment of inertia per unit mass,
$- W$ = exhaustion of gravitational energy.

The first entry in Table 2.6, (p. 74), for $m_c = 0 \cdot 0^{10}1 \times 10^{27}$ g, corresponding to only a minute central liquid core, happened to be more convenient to compute by the same programme than would have been $m_c = 0$ precisely, but may be taken as

Table 2.6. Three-zone models of the Earth

m_0 (10^{27} g)	ρ_0 (g cm^{-3})	p_0 (10^{12} dyn cm^{-2})	r_0 (km)	p_{12} (10^{12} dyn cm^{-2})	r_m (km)	p_{23} (10^{12} dyn cm^{-2})	m_{12} (10^{27} g)	r_s (km)	f	fr_s^2 (10^{18} cm^2)	$-W$ (10^{40} erg)	ρ_m (g cm^{-3})
0·0101	10·06	1·634	1·3	1·634	6473	0·0800	5·474	6741	0·3783	0·1719	0·2180	4·656
0·1	10·44	1·910	1327	1·650	6445	0·0833	5·460	6723	0·3752	0·1696	0·2196	4·695
0·2	10·64	2·068	1667	1·647	6417	0·0865	5·447	6704	0·3723	0·1673	0·2212	4·735
0·3	10·80	2·200	1905	1·640	6390	0·0898	5·434	6685	0·3695	0·1651	0·2228	4·776
0·3708	—	—	2042	1·634	—	—	—	6671	—	—	0·2240	—
0·4	10·94	2·317	2093	1·631	6362	0·0930	5·422	6666	0·3667	0·1630	0·2244	4·817
1·0	11·53	2·865	2825	1·539	6198	0·1125	5·359	6549	0·3522	0·1510	0·2339	5·079
1·876	12·09	3·448	3477	1·348	5961	0·1410	5·295	6371	0·3358	0·1363	0·2468	5·516
3·0	12·55	3·991	4076	1·040	5661	0·1775	5·257	6129	0·3238	0·1216	0·2611	6·196
5·0	12·98	4·532	4925	0·367	5131	0·2425	5·292	5654	0·3333	0·1066	0·2774	7·892

Earth mass = 5.976×10^{27} g $b = 3.5$ throughout $p_{23} = 0.08003 + 0.0325\,m_0$

Liquid $a = 1.3416 \times 10^{12}$ dyn cm^{-2}, core: $\rho_u = 6.260$ g cm^{-3}.

Mantle: $a = 2.1505 \times 10^{12}$ dyn cm^{-2}, $\rho_u = 3.975$ g cm^{-3}.

Outer $a = 1.1673 \times 10^{12}$ dyn cm^{-2}, shell: $\rho_u = 3.298$ g cm^{-3}.

representative of an initially all-solid Earth. Its overall radius is slightly greater than the corresponding model of Table 2.4 (p. 66) owing to the smaller value adopted for the uncompressed density of the mantle region, and implies a total contraction of about 370 km; so this possibly more accurate calculation leads to an even greater amount of contraction. It is seen that as the mass of the liquid core increases, the overall radius r_s decreases, and in fact it does so more or less linearly with the value of m_c.

An important feature of all these calculations is that it is the *changes* in the various quantities, as between one value of m_c and another, in particular $m_c = 0$ and $m_c = 1.876 \times 10^{27}$ g, the present core-mass, that are of main interest. It is found that if the calculations are carried out on some other basis, as for example two-zone models consisting solely of liquid core plus mantle (with no outer-shell zone), these differences are almost unaltered. For example, the values of W for $m_c = 0$ and 1.876, differ by 2.88×10^{38} ergs for the three-zone models, and by 2.86×10^{38} ergs for the two-zone models. This figure represents the total amount of gravitational energy that would have been released through contraction during the whole age of the Earth, and is as much as 4.8×10^{10} erg g^{-1}. It is of interest that this is of just the same order as the total amount of radioactive energy released, computed on the basis of modern estimates of the proportions of uranium, thorium, and potassium—the elements concerned—likely to have been present in the primitive planetary material. Gravitational energy of contraction will thus in part at least augment the radioactive energy, and once contraction has begun, cause the internal temperatures to rise more rapidly than would otherwise occur in the absence of contraction. It needs the presence of radioactive materials for this further source of energy to be switched on, as it were, but some of it might be stored as elastic energy and some would go into energy of ionization in the core, leaving only a part actually to increase the temperature.

A second interesting feature of Table 2.6 (p. 74) is the decrease of moment of inertia, Mfr_s^2, with the increase of core-mass.

Initially this would have been about $1\cdot72 \times 10^{17}$ cm² per unit mass, whereas at the present time it has decreased to $1\cdot36 \times 10^{17}$ cm² per unit mass, values that are roughly in the ratio 5:4. This means that if the rotatory speed of the Earth were a purely intrinsic matter, unaffected by the moon and sun, the present rotation period of twenty-four hours would have required an initial period of thirty hours. This decrease of period is in fact comparable with though somewhat less than the increase believed to have resulted from tidal friction in the whole age of the Earth. But the theory of tidal friction, which at present suggests an embarrassingly short period for the time for which the moon can have been a satellite of the Earth, has not yet taken into account the possibility of an additional factor of this internal kind.

Another feature of Table 2.6, and one that came as a considerable surprise, arises from the first few values of the interface-pressure, p_{12}, between liquid core and mantle for small values of the core-mass m_c. Whereas every other quantity in this table (except f) either always increases or always decreases as the core-mass increases, this interface-pressure at first *increases* to a maximum and thereafter decreases, reaching the original central pressure again when $m_c = 0\cdot3708$, and falling eventually to zero when the core-mass is equal to that of the whole Earth. It has been seen (Fig. 2.3, p. 71) that, with temperature produced by radioactive heating, most of the Earth is practically isothermal and certainly the central regions are; so that if the liquid core is a phase-change produced by temperature and high pressure, this increasing interface-pressure would mean that, once the tiniest liquid core had formed, the lowest part of the mantle would be at a higher pressure than was originally needed at the centre for the phase-change but at the same temperature. It therefore would also undergo the phase-change, without requiring any further radioactive heating, and so on, and the process could cease only when the interface pressure had fallen below the original central value, which as Table 2.6 shows would mean when m_c had exceeded $0\cdot3708 \times 10^{27}$ g, or just over

6 per cent of the entire mass of the Earth. The phase-change here involved probably partakes of the nature of collapse of the outer electron-shells, and as the requisite pressure arises gravitationally it will be communicated as rapidly as the collapse can occur, which would mean in a matter of minutes! Here the motion is being followed by means of a series of static models, but it is plain that in reality the collapse would overshoot the mark, and that possibly a somewhat greater mass than $m_c = 0.3708 \times 10^{27}$ g would be liquefied. The radius of this initial core, despite its small mass, exceeds 2000 km (Table 2.6, p. 74), which means nearly one-third the initial radius of the whole planet. Because of the great *increase* of density associated with the phase-change to liquid form, the material of the outermost part of the core would have fallen in by as much as 460 km from the position it occupied in the former all-solid model. The amount of radial contraction outside the core becomes progressively less than this, but the difference between the initial value of r_s (for $m_c = 0$) and the final value (for $m_c = 0.3708$) shows that the surface contraction would be as much as 70 km, implying a decrease of superficial area of almost 12×10^6 square kilometres.

The release of energy accompanying such a gravitational collapse is stupendous by any ordinary standards; it is seen (from Table 2.6, column 12) to amount to about 6×10^{37} ergs. The largest earthquake yet recorded had a total energy of about 3×10^{27} ergs, so that in the few minutes of this collapse a release of about 2×10^{10} this amount of energy would have occurred. Had there been any inhabitants of the Earth at that time, such an event is unlikely to have passed without notice. Fortunately, once a core-mass in excess of this critical value (0.3708×10^{27} g) is reached, the interface-pressure continues to decrease and the planet becomes thoroughly stable against any further rapid collapse, though radioactive heating will continue to bring about steady increase of the core-mass so long as the release exceeds the gentle loss of heat-energy arising from the very slow processes of conduction within the Earth. This means that as far as subsequent gradual contraction is concerned, a further decrease of radius of about 300 km has since occurred, implying a decrease of circumference by about 1900 kilometres and of surface area by about 54 million square kilometres.

The details of how a collapsing planet would behave under a world-wide process of this kind defy imagination. As noted above, it may be conjectured that the collapse would continue beyond the state corresponding to core-mass $m_c = 0.3708 \times 10^{27}$ g until sufficient recoil was built up to reverse the motion, and then presumably an outward-travelling compression-wave would move through the planet. On reaching the free surface such a wave might have most interesting effects: for instance, there is the possibility that as the surface is approached a great deal of energy of radial motion might be fed into a comparatively thin layer, and something of the nature of a whip-crack effect might occur, though here of course in purely radial motion. If the outward speed attained should exceed the escape velocity, the Earth might even shed into space a surface layer of material, already much fractured one would suppose by the preceding collapse if not also by earlier general expansion. This possibility, though admittedly highly speculative, may not be so absurd as it might seem at first sight: for even if only 1 per cent of the released gravitational energy eventually went into systematic motion with escape speed, this would nevertheless involve a mass of about 10^{24} g, which is of order of 1/10 000th the mass of the whole Earth. The possibility that some of the meteoritic material of the solar system may have been projected into interplanetary space in such a way, perhaps through the collapse of more than one planet, would make a theoretical study of this possibility of much interest, but it would obviously present a problem of great mathematical difficulty.

It is equally hard to see how the surface layers would react to the initial inward collapse. As has already been seen, during the long quiescent stage while the central temperature was rising to the critical value for liquefaction, the outer-shell zone would probably have increased in depth and the overall radius *increased* by a few kilometres; consequently, some rifting of the surface layers might already have taken place. An increase of radius of only 6 km would lead to an increase of surface area of about 1 million km^2; so the initial collapse might have found the outer layers already considerably sundered by lateral stretching and in a state suitable for the thrusting of one layer over another. Moreover it is possible, as has already been seen, that material lower down, at a depth of a few hundred kilometres or less, might already have become melted through radioactive heating; if so, fissures of the superincumbent layers might have permitted the rising to the surface of lighter melted materials through ordinary hydrostatic forces. It could even be that the observed lower-density materials of the extreme outer layers, in particular those of the continents, represent one consequence of such processes.

The account of the evolution of an initially all-solid Earth that the phase-change hypothesis for the liquid core drives us to is therefore seen to be almost a complete reversal of the classical theory based on the notion that the planet began as an entirely molten mass. In the older theory, radioactivity, although considered capable of producing minor heating effects, was not regarded as playing any important role in determining evolution, whereas on the theory here presented it is all-important. The liquid core was ascribed to the sinking of free heavy metals, principally iron, to the centre, the molten state meanwhile persisting throughout the greater part of the Earth. Differential contraction of the outer layers as cooling and solidification penetrated downwards was proposed as the cause of mountain-building, as earlier explained.

Acceptance of this seemingly inevitable scheme was for many years fairly general but has gradually diminished over the past two or three decades for various reasons. As has already been mentioned, the possibility of a body of so small a mass as the Earth condensing quickly from high-temperature material has come to be regarded with increasing doubt, and it is now considered more likely that the Earth accumulated comparatively gradually from a dust-distribution associated with the sun in such a way as to have been initially solid throughout. Again, the inadequacy of thermal contraction to meet the geologically required crustal shortening could also be urged. Moreover, if the core consisted largely of iron, it would imply a most unusual composition for the Earth.

On the other hand, if the idea of an all-solid initial state is adopted, the main basis of the older theory would thereby be removed, and it would become necessary to construct a new picture of terrestrial evolution. The only available driving-force would appear to be heating by radioactive energy, and this leads on to the notion that the liquid core represents a phase-change. Since a metallic state is almost certainly required, together with a much higher 'uncompressed' density, the change probably involves ionization and liquefaction of the material as a result of both pressure and temperature. By its nature, little can be known at present of such a process, since high-pressure experiments with static pressures have so far been limited to the pressures and temperatures that the materials of any apparatus can withstand. Shock-wave experiments have produced high pressures comparable to those of the Earth's deep interior, but to what extent the results can be trusted as indicative of static pressures remains highly uncertain. Moreover, such experiments, more or less of necessity have not been conducted with materials initially at the relevant temperatures of some thousands of degrees.

By way of conclusion, the possible course of evolution of the Earth may be summarized in brief as follows: The planet began its independent existence as an entirely solid body. Rise of internal temperature would gradually occur through release of radioactive energy. The level of discontinuity between mantle and outer shell would move towards the centre, and acting alone would lead to slight expansion. Since in the core–mantle phase-change the liquid form has higher density, the melting-point would have a minimum at the centre of the planet and this phase-change would first commence there. The resulting density increase would be sufficiently great to render the planet catastrophically unstable, and some 6 per cent of the mass would liquefy extremely rapidly, accompanied by collapse of the whole planet. With this stage past, the planet would become stable, and increase of the liquid core would proceed only gradually as the temperature continued to increase, accompanied by slow contraction of the Earth as a whole. This would build up increasing stresses in the outer layers until they became so great that the strengths of the materials were exceeded, and at such places folding and thrusting would occur over a comparatively short time with a rapid decay to a negligible rate of surface contraction. These eras of mountain-building may be mainly confined to periods of time shorter than the intervals that separate them. In these intervals, rising temperature would continue to cause the external region to expand, though the total amount of such expansion would be of far smaller order than the overall contraction.

The extent to which such a picture may prove correct is a matter for the future. Even at present the theory suggests that the Earth is a more complex object than was formerly thought, and of course it is this very complexity that gives it such remarkable properties. To establish or disprove the phase-change hypothesis will require a greater understanding of high-pressure physics than exists at present, but it is of interest that the theory points the way to a realm where inquiry is urgently needed; if it does no more than this it will have served some scientific purpose. Meanwhile, we must continue to regard the interior of the Earth as a great mystery, for it is certain that much remains to be discovered about it.

3 The Constitution of the Terrestrial Planets

THE planets of the solar system can be divided into two distinct groups. One consists of the four great outer planets: Jupiter, Saturn, Uranus, and Neptune, which, as their low densities show, must almost certainly be largely composed of the lightest gases. The other group consists of the four small high-density inner planets: Mercury, Venus, the Earth, and Mars, together possibly with the moon (see Table 2.1, p. 46). It is with these last five bodies and their possible relationship that the present chapter is concerned. (The status of Pluto can scarcely be settled as yet because of the large uncertainties attaching to both its measured mass and dimensions, though present estimates of the mass would rank it as an object similar to members of the terrestrial group.) At first sight it seems surprising that these small inner planets should have densities so much greater than the much more massive outer ones. (Uranus, the smallest, is equal to about 15 Earth-masses, and Jupiter, the largest, about 318 Earth-masses.) As has been seen, however, this probably arises solely from their particular locations in the solar system. Because of the temperature resulting from its proximity to the sun, the Earth cannot retain a hydrogen atmosphere, and so could not increase in mass from a gaseous nebula surrounding the sun and consisting mainly of hydrogen. The same holds even more strongly for the other terrestrial planets and the moon. However, if a planet similar to the Earth were situated at the distance of Jupiter or

Saturn from the sun, the temperature there would be so much lower that hydrogen could be retained, and a small planet could therefore increase in mass by acquisition of material within a gaseous nebula. If, as seems probable, all planets begin as accumulations within a disk of dust, then the four great outer planets may have small central cores of material of terrestrial kind. With the enormous central pressures obtaining, the physical state of this material can scarcely be even vaguely conjectured.

The present atmosphere of the Earth, although of such great importance to ourselves, contributes only about one-millionth of the entire mass of the planet. For Mars, Mercury, and the moon, the proportions by mass are all far less. The extent and mass of the atmosphere of Venus are at present only very imperfectly known, but even if the mass were as high as 10 or 100 times that of the Earth's atmosphere, it would still be a negligible part of the whole mass of the planet. So the central problem of these planetary neighbours concerns in the first place the compositions of their interiors. If we rely on a general principle of uniformity in nature, the simplest hypothesis to consider first is that all the planets formed initially of similar materials. We may then ask whether this hypothesis can be sustained in the face of their different mean densities and other properties. Clearly it is a matter of prime importance where the origin of the system is concerned to have some answer to this question. If the planets *are* all of the same composition as the Earth, then it is seen at once how some progress can be made towards understanding their interiors: the relevant elastic properties at any particular pressures within them can be supposed to be the same as within the Earth at those pressures, and given the mass of a planet it would then be possible to proceed as for the Earth, and calculate the internal distribution of density and the overall radius and other related properties.

Accordingly the most fundamental property of a planet that it is necessary to know is its total mass. This is readily found with considerable accuracy if a planet possesses a satellite or satellites, as of course is the case for the Earth and for Mars; but the planetary theory does not lead to values of sufficient accuracy for either Mercury or Venus, and gravitational experiments are required for both these planets. Such an experiment has already been successfully carried out for Venus by the fly-past of the space-vehicle Mariner II. This led to a value of 4.86954×10^{27} g on the basis that the mass of the Earth is 5.976×10^{27} g. This means that Venus has 0.81485 times the mass of the Earth, and in this respect is therefore almost a twin body. But it has to be remembered that the quantity that emerges from the trajectory calculations is not the actual mass but the product GM_V of the mass of Venus multiplied by the constant of gravitation, G, compared with the similar product GM_E for the Earth, and the quantity G is itself not known to much better than 1 part in 1000. (The best determination of it at present is $6.670 \pm 0.005 \times 10^{-8}$ cgs units.) Fortunately this constant enters the expression for the theoretical radius of a planet only as $G^{1/4}$, so that error of the order of only 1 km would result from this uncertainty. (There is an urgent need for a method of measuring G to a higher order of accuracy, but it is very difficult to devise an appropriate experiment.)

So far as Mercury is concerned, apart from the possibility of a chance close encounter with a comet, its mass can as yet be found only from the perturbations it produces on the other terrestrial planets, and these are so small that the mass is not certain to within ± 10 per cent of any particular value obtained. As there is also considerable uncertainty in the measured radius, of at least ± 5 per cent, the calculated mean density is subject to large probable error, so large in fact that it is doubtful to what extent it is yet worth while attempting any inferences about the internal structure of the planet. A flight close to Mercury by a space-vehicle is recognized as one of the most urgent requirements of planetary physics, and fortunately a mission is now under active consideration. The ideal experiment, of course, would be for artificial satellites to be placed into orbit round both Venus and Mercury, and this no doubt will eventually be accomplished. The mass of the moon, on the other hand, is known with high accuracy from the direct gravitational effects it produces on the Earth.

Equally important to our purpose is a knowledge of the radius of the solid surface of the planets. The surface of Venus seems always to be entirely hidden by impenetrable cloud cover, and the theory of the structure of the atmosphere of Venus remains too indefinite so far to supply an accurate value of the depth of the solid surface below the observed optical surface. However, the solid surface of Mars is clearly visible, and while different methods of measurement lead to different values of the radius, modern results, including that of the Mariner IV fly-past of July 1965, combine to indicate a value not much below 3400 km. The same fly-past also gave an improved value of the mass of Mars, reducing it slightly below the value derived from planetary theory and the motions of its satellites by about 0·23 per cent to $0·6423 \times 10^{27}$ g, or 0·10748 times the mass of the Earth.

There is a third less obvious quantity theoretically associated with any almost spherical mass-distribution that provides an overall indication of the extent to which it is condensed towards its centre. For a perfectly uniform sphere of mass M and radius R, the moment of inertia about a diameter is fMR^2 with $f = 0·4$, but with inward increase of density the numerical factor f becomes less than 0·4, and progressively so as concentration towards the centre increases. For example, for the Earth at present the value is 0·3335, while for the moon it is believed to be 0·3971, which would mean only very slight departure from uniform density throughout. Now for a curious mathematical reason it happens that f is simply related to the flattening of a slowly rotating oblate planet in hydrostatic equilibrium (the flattening is defined as the amount by which the ratio of the polar radius to the equatorial radius falls short of unity), and the orbit of a satellite is influenced in easily determinable ways by any departure of the controlling planet from strict sphericity. Accordingly, if a planet possesses a satellite or satellites, analysis of their observed motion may afford a measure of the internal distribution of density by means of this factor f, and the closer to the planet the satellites move the more pronounced are the relevant effects. This is yet another reason why artificial satellites in orbits close to Mars, Mercury, and Venus—and indeed almost any other astronomical body—would supply important information. The satellites of Mars indicate for this flattening, or the so-called dynamical ellipticity, a value of $1/191 \cdot 8$. For the Earth, the lunar orbit had already yielded a value of $1/297 \cdot 1$, but close artificial satellites have led to an improved value of $1/298 \cdot 2$ with much higher degree of accuracy.

This moment-of-inertia factor f cannot be measured for either Mercury or Venus at present in the absence of any satellite, as already explained, but even if in time these two planets were supplied with artificial satellites, they are both rotating so slowly that their resulting departure from spherical shape, which depends on the square of the angular velocity, might be so minute as to be difficult to disentangle from the effect of other irregularities of shape. So it may be only by much more indirect theoretical means that the degree of central condensation within these planets is likely to be found.

With actual exploration of other planets becoming more and more feasible, the degree to which they resemble the Earth in other respects than mere size and mass can also be investigated. For example, the existence or absence of magnetic fields, already considered to be non-existent on the moon and Mars, or at least immeasurably small by fly-past equipment, may provide an indication of whether a planet possesses a fluid metallic core. It is within such a core that the main magnetic field of the Earth is believed to originate, though the precise mechanism of its production is by no means yet thoroughly understood. Then again, the existence or otherwise of folded and thrusted mountains, which are already thought to be absent from the moon, may enable inferences to be made about the internal structure and evolution of other planets. It would clearly be of the greatest interest and importance to know whether moonquakes, marsquakes, and so on, occur on these other bodies, and if so with what degree of intensity and frequency. If any planets do undergo such seismic disturbances, it may in the distant future be possible to investigate their internal properties by methods similar to those already applied to the Earth. Meanwhile answers to such questions can be attempted only in a very preliminary way by theoretical means, and with this in mind we turn to the question to what extent the structures of the other terrestrial planets are consistent with the hypothesis that their general compositions are similar to that of the Earth.

VENUS

Beginning with Venus, whose mass is most nearly comparable with that of the Earth, we can make calculations exactly similar to those made for the Earth. In the first place it is readily shown that the central pressure in Venus must at the present time *exceed* $1 \cdot 34 \times 10^{12}$ dyn cm^{-2}. As the pressure at the core–mantle boundary in the Earth is $1 \cdot 37 \times 10^{12}$ dyn cm^{-2}, this immediately suggests that Venus is to be regarded as having three main zones, like the Earth, including a liquid core. But if the Earth began as an all-solid body, it is even more likely that the somewhat less massive Venus was also solid throughout when it commenced its existence. Looking back to this stage, the first question that arises is what the initial radius in such a form would be. Exactly analogous calculations for this can be made on the basis that Venus would then have possessed a central region consisting of mantle-type material together with an outer shell. The mass of the central region can be given a succession of values to yield a series of models, and the initial value of the interface-pressure, at the outer boundary of the mantle material, would be the parameter determining the model relevant to this stage of the planet. As for the initial Earth, the precise value of this pressure is at

7—M.S.S.

present not known except that it would almost certainly be less than the present terrestrial value, 0.141×10^{12} dyn cm^{-2}. The results of such calculations are shown in Table 3.1.

Table 3.1. All-solid models of Venus

m_c $(10^{27}$ g)	ρ_c (g cm^{-3})	p_c $(10^{12}$ dyn cm$^{-2})$	r_c (km)	p_{int} $(10^{12}$ dyn cm$^{-2})$	r_s (km)	ρ_m (g cm^{-3})
4·0	5·706	1·413	5795	0·1510	6335	4·573
4·1	5·707	1·415	5849	0·1332	6326	4·592
4·2	5·708	1·416	5903	0·1155	6317	4·612
4·3	5·709	1·417	5956	0·0980	6308	4·633
4·4	5·710	1·418	6008	0·0806	6298	4·654
4·5	5·710	1·418	6060	0·0632	6288	4·676
4·86594	5·711	1·419	6248	0·0	6248	4·766

 m_c = mass of material in mantle form (in central region)
 ρ_c = central density
 p_c = central pressure
 r_c = radius of central mantle zone
 p_{int} = pressure at interface between mantle and outer shell
 r_s = overall radius to outer surface
 ρ_m = mean-density of whole model

Table 3.1 shows that only for values of m_c in excess of 4.0×10^{27} g, that is with more than 80 per cent of the mass in mantle form, does the interface-pressure fall below 0.141×10^{12} dyn cm^{-2}. It can therefore be inferred at once that only such models are relevant to an initial all-solid form for Venus. As for the Earth, it is seen that ρ_c and p_c change very little over the whole range of m_c shown, and the mean density ρ_m changes not much more. For the Earth, in computing three-zone models, it was supposed that p_{int} was initially about 0.08×10^{12} dyn cm^{-2}, but Venus, perhaps forming nearer to the sun (though we cannot really be certain of this merely because it is so situated now) might have a corresponding initial p_{int} somewhat higher because of the higher local temperature. For this reason we might adopt the value 0.098×10^{12} dyn cm^{-2} corresponding to $m_c = 4.3 \times 10^{27}$ g. However, it is not in fact essential for present purposes to make any precise decision on this point, for the currently observed radius of Venus to the top of the cloud-layer is actually 6200 km, whereas every value of r_s in

Table 3.1 substantially exceeds this figure. Accordingly it may be concluded that the present Venus, if its composition is similar to that of the Earth, cannot possibly be an all-solid body. We are led to the clear possibility that Venus possesses a central liquid core, and in addition will have undergone contraction to whatever its present actual solid radius may be.

An important urgent requirement of planetary science is some direct means of determining this solid-surface radius of Venus; unfortunately, theories of the atmosphere do not as yet determine accurately the depth of the surface below the observed optical boundary. If a depth of 75 km is adopted, the solid radius would be 6125 km, and on this basis calculations can be made of the amount of contraction of radius and surface area implied by the foregoing models. The results are shown in Table 3.2.

Table 3.2. Contraction of Venus

Decrease of r_s and of surface-area, and depth of outer shell

($R_V = 6125$ km adopted as present solid radius)

m_c (10^{27} g)	p_{int} (10^{12} dyn cm^{-2})	$r_s - R_V$ (km)	Decrease of surface area (10^6 km^2)	Depth of outer shell $r_s - R_V$ (km)
4·0	0·151	210	32·9	540
4·1	0·133	201	31·4	477
4·2	0·116	192	30·0	414
4·3	0·098	183	28·5	352
4·4	0·081	173	27·0	290
4·5	0·063	163	25·4	228

Table 3.2 shows that, unless the depth of the atmosphere is very different from that supposed, the amount of radial contraction can have been only about two-thirds that for the Earth, and the corresponding surface contraction only about one-half, though nevertheless quite impressive at some 30×10^6 km^2. It is seen, however, from Table 3.1 (p. 88) that the central pressure in an all-solid Venus would have been about $1·42 \times 10^{12}$ dyn cm^{-2}, which is almost one-eighth *less* than in an initially all-solid Earth. Accordingly, on the phase-change hypothesis for the nature of the liquid core, a *higher*

central temperature, and presumably to a comparable proportional extent, would have been needed to produce conditions
in Venus suitable for the formation of a liquid core. This would
mean a temperature a few hundred degrees higher than at the
same stage in the development of the Earth. Despite the
possibility of a slightly higher initial internal temperature of
the newly formed Venus as compared with the Earth (because
of its possible closer proximity to the sun at this time), this
lower central pressure in turn might well mean that Venus
would have required a time approaching 10^9 years longer to
reach the equivalent stage than did the Earth. For this reason,
Venus may be less far along its course of internal evolution
than the Earth. Thus the foregoing amount of contraction may
have been spread over a rather shorter length of time than has
been that for the Earth. At all events, these considerations
strongly suggest that Venus may possess folded and thrusted
mountains at its outer solid surface resembling those of
terrestrial type.

Calculations of models of Venus with the mass of the liquid
core as parameter can be carried out in exactly the same way
as for the Earth (described in Chapter 2), and closely similar
results emerge that need not therefore be given in detail here.
For example, the interface pressure between liquid core and
solid mantle begins by increasing to a maximum and thereafter
decreases steadily to zero as m_c increases from zero up to the
whole mass of the planet. As for the Earth, this feature of
initial increase would imply a collapse of the planet once the
critical central temperature for melting had been reached. The
core mass (of the static model) for which the interface pressure
has fallen again to the original all-solid central pressure amounts
to about 4·8 per cent of the whole mass of Venus, and the liquid
region has a radius about one-third that of the whole planet.
The resulting collapse at the outer surface would be by about
52 km, and the sudden release of gravitational energy would be
about 3×10^{37} erg, or about half the computed amount in the
case of the Earth. All these figures would be somewhat altered if

the adopted solid radius were different from 6125 km, though by no means would they be altered in order of magnitude.

As has been mentioned, such a collapse would probably overshoot the configuration determined on the basis of static models, but once equilibrium became restored the planet would henceforth be thoroughly stable against any further collapse; subsequent growth of the liquid core would proceed quite steadily as the internal temperature continued to rise, as described in Chapter 2 for the Earth. Thus, if the core-mass has increased substantially beyond this first stage, it can be expected that mountain-systems will have formed on Venus comparable with those on Earth. Calculations show that the present core-mass, to agree with a surface radius of 6125 km, would be about $1 \cdot 0 \times 10^{27}$ g, about one-fifth the entire mass, and the core-radius just under half the overall radius of the planet. These figures are necessarily subject to some uncertainty because of the lack of definite knowledge of the present solid radius, and in addition to the need for accurate measurement of the radius there is also the related problem of the nature and extent of any surface irregularities on Venus.

The existence of a liquid metallic core within Venus would suggest the possibility of a magnetic field for the planet. But the production of such fields probably depends also upon the planet possessing rotation, and may also be associated with precession of the planet. For Venus, the rotation rate is extremely slow by other planetary standards: the latest radio measurements are interpreted to indicate a period longer than the orbital period of 225 days, and with direction of rotation roughly the opposite of this. Any magnetic field might accordingly be expected to be correspondingly small as compared with that of the Earth, though perhaps not entirely absent. It would therefore be highly interesting to try to measure any such field by delicate instruments landed on the surface to minimize the inverse-cube effect. These might provide some indication of whether the field-strength were in general proportion to the angular rate, since some difference might result possibly through the smaller size of the central core, or whether it were of a lower order altogether, suggesting some different degree of dependence on the angular velocity.

MARS

Of the planets in the solar system, Mars is the most favourably situated for observational study, not only through its proximity but because its orbit lies outside that of the Earth. Its thin transparent atmosphere permits details of the solid surface and of cloud formations to be seen. In addition, Mars is in rotation at a rate very similar to that of the Earth (the Martian day is 24 h 37 min) and it possesses two tiny close satellites. From a dynamical point of view Mars provides a much more complete and versatile system than does Venus. The red planet, as it is sometimes termed, has long been the subject of considerable speculation, not all of it well founded, as possibly sufficiently similar to the Earth to represent a suitable abode for life. But with growing knowledge of its atmospheric conditions, the likelihood of this entertaining possibility has dwindled somewhat. Here, however, we shall be concerned with the interior of the planet, and on the basic hypothesis that its composition is similar to that of the Earth it nevertheless turns out to have very different internal structure. This arises solely from the much smaller mass, which is just under one-ninth that of the Earth.

A lower limit to the central pressure in Mars can readily be estimated, for there are general arguments showing that it must exceed the value that would obtain if the density were perfectly uniform; for this case the central pressure would in fact be about 0.25×10^{12} dyn cm^{-2}. Furthermore, unless the material were strongly concentrated at the centre at much higher density, the value could not greatly exceed this. It can accordingly be settled fairly definitely at the outset, and detailed later calculations confirm this, that the central pressure in Mars is less than one-tenth the value at the centre of the Earth. In fact, this greatest pressure in Mars is exceeded within the

Earth at a depth of a mere 700 km, and it is known from the seismic data that the Earth is solid at this depth. It is also possible, for two reasons, that the Earth is at higher temperature at such a depth than is material near the centre of Mars. First, if the material from which a planet accumulated was originally at much the same distance from the sun as Mars moves now, then the newly formed Mars would have commenced its existence at a temperature at least some 50–100°K less than that of the Earth because of the different distance from the sun. Secondly, Mars being a much smaller body than the Earth (it has only just over half the radius), the escape of radioactively produced energy would occur more quickly mass for mass. This implies that although the central temperature in Mars could be almost as high as for the Earth, the latter would still be almost iso-thermal to a distance far beyond the boundary of the liquid core and perhaps remain so as far out as within 700 km of the surface, whereas in Mars the temperature will have begun to decrease outwards rapidly at a greater depth than where this occurs in the Earth. But the Earth remains solid down to a depth of 2898 km, where the pressure is $1\cdot37 \times 10^{12}$ dyn cm^{-2}, nearly five times the maximum that occurs in Mars. Internal tempera-tures in Mars would therefore need to be far *higher* than in the Earth to bring about the phase-change to liquid form. Accor-dingly on this hypothesis as to the nature of the terrestrial liquid core, and the general hypothesis of uniformity of com-position, it seems reasonably certain that the planet Mars must be solid throughout.

In order to test this conclusion, it is necessary to calculate theoretically some quantity associated with the planet that can be compared with observation. Clearly the most straightforward thing to compute for this purpose would be the overall radius. If regard is had to all direct observational determinations, there is considerable uncertainty. Russell, Dugan, and Stewart, for example give the value $3396\cdot5 \pm 8\cdot0$ km, while de Vaucouleurs, from a discussion of almost all available measures, obtains a mean value of $3408\cdot8 \pm 3\cdot6$ km. As the optical flattening, determined from the observed elliptical shape of the disk of the planet, is believed to be equal to about 0·013, the polar diameter of Mars must be about 44 km less than this, and the equivalent spherical radius would be 3394·0 ± 3·6 km. The latest determination was made by the recent fly-past of Mariner IV, making use of the occultations of the spacecraft by the planet. This led to a value of 3390 ± 6 km. It would be desirable for comparison with theory to know the equivalent spherical radius to within better than 1 km; this again is something that must be a principal objective of future missions to Mars. There is always the possibility that any particular method may be subject to undisclosed systematic error, and it is not in general possible to assess such kinds of error until some new method is devised that attains superior accuracy by at least a whole order of magnitude. Meanwhile, where Mars is concerned, it will simply remain impossible to regard the radius as known to within at least ± 10 km.

On the view that Mars is solid throughout, it is plain that only two-zone models will be required to determine the internal distribution, the overall radius, and other related properties. The procedure is thus exactly similar to that already considered for computing all-solid two-zone models of the Earth and Venus. Table 3.3 (p. 95) shows the results of a series of values of m_c, the mass of material in mantle form of the central region. Since the dynamical ellipticity of Mars can be determined from the satellite motions, it is relevant here to include also the moment-of-inertia factor f, defined as such that the moment of inertia is fMr_s^2, and which is related to the dynamical ellipticity by the so-called Radau formula.

Column 6 of Table 3.3 shows the overall surface radius r_s for the listed series of models, and closest agreement with the above-quoted observed values evidently occurs for $m_c = 0·4 \times 10^{27}$ g, for which $r_s = 3398$ km, though in view of the uncertainty in the measured radius the appropriate value of m_c is probably covered by the range 0·36–0·42 × 10^{27} g. It is also seen that over this range both the central density and the

Table 3.3. All-solid two-zone models of Mars

m_c (10^27 g)	ρ_0 (g cm^-3)	p_0 (10^12 dyn cm^-2)	r_0 (km)	p_{int} (10^12 dyn cm^-2)	r_s (km)	f	ρ_m (g cm^-3)	$r_s - r_0$ (km)
0·1	4·484	0·2574	1756	0·1727	3493	0·3863	3·603	1737
0·2	4·502	0·2699	2217	0·1357	3463	0·3821	3·698	1246
0·3	4·514	0·2780	2543	0·1025	3431	0·3803	3·802	888
0·32	4·515	0·2792	2600	0·0961	3423	0·3802	3·824	823
0·34	4·517	0·2803	2654	0·0898	3418	0·3803	3·846	764
0·36	4·518	0·2818	2706	0·0836	3411	0·3804	3·869	705
0·38	4·520	0·2824	2757	0·0775	3404	0·3807	3·893	647
0·4	4·521	0·2833	2805	0·0714	3398	0·3810	3·916	593
0·42	4·522	0·2841	2853	0·0653	3391	0·3815	3·941	538
0·44	4·523	0·2849	2898	0·0593	3383	0·3820	3·966	485
0·46	4·524	0·2855	2943	0·0533	3376	0·3827	3·991	433

pressure change very little. The fact that the latter only slightly exceeds the value for uniform density (0.25×10^{12} dyn cm^{-2}) is one indication that Mars must have fairly uniform density itself, and this is also shown by the fact that the central density of 4.52 g cm^{-3} does not much exceed the mean density of 3.92 g cm^{-3} (for $m_c = 0.4$).

For this range of m_c, the pressure at the interface between mantle and outer-shell material changes from 0.084 to 0.065×10^{12} dyn cm^{-2}. Accordingly, almost whatever the actual radius is eventually found to be, this pressure can be only about one-half that at which this discontinuity occurs in the Earth at present, which is 0.141×10^{12} dyn cm^{-2}. This result, however, is perfectly consistent with the view that the temperature within Mars at the depth of the interface (about 600 km) is likely to be less than the temperature within the Earth at the discontinuity, and that as a consequence a lower pressure is needed to effect the phase-change to the denser crystal form involved in the difference between outer-shell and mantle-type material.

Since Mars probably began its existence with a lower internal temperature regime than the present one, it may be conjectured that the initial interface pressure was lower still. As Table 3.3 shows, this would mean that the relevant model would correspond to a higher initial value of m_c and therefore a somewhat smaller total radius r_s. This in turn would have the interesting consequence that the radius would in fact *increase* slightly as the temperature rose and the required interface pressure increased. If, to select some values simply for numerical illustration, the initial and present values of this pressure were 0.053×10^{12} dyn cm^{-2} and 0.065×10^{12} dyn cm^{-2}, corresponding to $m_c = 0.44$ and $m_c = 0.40$, the associated increase of radius would be about 15 km, and the accompanying *increase of surface area* just over 1 million km^2. If this has been the general trend of evolution of Mars, it is possible that such expansion would result in considerable rifting of the outer layers. But because the central pressure is too low for a liquid core to form, the planet will not have undergone any contraction but only this expansion, and no systems of folded and thrusted mountains would therefore be expected at its surface. It is not at all clear on what scale such rifting would take place, whether causing a few wide channels of lengths comparable with the radius of Mars and dividing the surface into a few areas, or a great many short narrow ones dividing the surface into a large number of areas. It could also happen that if ordinary melting occurred in Mars at a few hundred kilometres depth through local radioactive heating or any other cause, as seems to have occurred within the Earth, the lighter liquid components could rise and reach the surface through these rifts, and perhaps more than fill them before spreading out sideways. It would be tempting to think of the much-disputed 'canals' in terms of such surface features, but until more is known of the nature of the Martian surface there seems little point in speculating on such possibilities.

Before the *Mariner IV* fly-past, observations of the surface could only be made telescopically, and only under the most perfect atmospheric conditions could Mars be seen much better than can the moon with the unaided eye. Even so, it had long been considered, from the absence of detectable shadows near the edge of the observed disk (where the sunlight would be falling nearly horizontally to make shadows longest), that there could be no localized surface irregularities of more than a few thousand feet in height and hence no really high mountains anywhere. The *Mariner IV* equipment photographed about 1 per cent of the entire area of Mars by a succession of about twenty more or less rectangular areas lined across a hemisphere of the planet. A full interpretation of the series has not yet been completed, but preliminary studies have not suggested that there are any indications of surface features that could correspond to mountain ranges. Indeed, the pictures reveal instead remarkable resemblance to the lunar surface as far as the more pronounced features are concerned.

The absence of a liquid core of metallic form in Mars leads to another conclusion of the highest interest. The rotation rate of the planet and the inclination of the axis are both almost the same as for the Earth, and other factors being the same any mechanism for producing a magnetic field by rotation should be equally effective within Mars. But the absence of a liquid core would render any such mechanism inoperative, and no main magnetic field would be expected. This prediction, which was in fact made long before *Mariner IV* set out on its seven-month flight to Mars, was in fact duly verified, or at least not contradicted, by the magnetometer measurements made close to the planet. No field strong enough to be detected by the equipment carried was found to exist, which means that the dipole-moment can be at most only 1/3000th that of the Earth. The argument is reversible, of course, now that measurements have been made, and from the non-existence of a magnetic field the absence of any sizable liquid core can be inferred.

If in fact the liquid core of the Earth were largely composed of iron drained out of the solid exterior, then for a similar initial composition of Mars a similar course of development would be expected. It is true that the average force of gravity within Mars is rather less than half the value in the Earth, but this is only a matter of degree, and would seem unlikely to prevent the supposed process of separation of the iron if conditions were otherwise the same. The process of liquefaction would be favoured in Mars, since the pressure at the centre is only about one-fifth that at the outer boundary of the Earth's liquid core, and hence the temperature required to melt iron deep within Mars would be only about one-half that needed within the Earth, if it is assumed that the change from solid to liquid involved is such that the solid form has higher density, as is the case for ordinary melting. Then again, if Mars possessed any substantial core of molten or even solid iron, this material is so much more compressible than mantle-material that the resulting overall radius would fall considerably short of the observed value. Calculations made on the basis that one-third of the mass is in the form of an iron core, the proportion indicated by the mass of the Earth's core, lead to a value no more than about 3200 km, about 200 km less than the observed value, and even on present evidence this seems quite unacceptably low. Though these arguments are perhaps not finally conclusive and inescapable, they nevertheless appear to provide strong indications that the planet can have no central zone similar to that of the Earth, and in fact that it cannot even possess an iron core in solid form.

As already mentioned, a second observational constraint is imposed where Mars is concerned by the value of the dynamical ellipticity as revealed by motions of its satellites. This circumstance enables a further comparison of the present theoretical results to be made. The values of the moment-of-inertia factor f and the mean density ρ_m in Table 3.3 supply one relation between these quantities, but a second relation having an entirely different basis arises from consideration of the slight departure from sphericity undergone by a slowly rotating planet in internal hydrostatic equilibrium. This is expressed by the well-known Radau formula, already referred to above, which may be written in the form

$$f \equiv \frac{I}{Mr_s^2} = \frac{2}{3} - \frac{4}{15}\left(\frac{15\pi}{2G\rho_m T^2 e} - 1\right)^{\frac{1}{2}}, \qquad (3.1)$$

where T is the period of rotation of the planet, accurately known for Mars, and e is the dynamical ellipticity. Accordingly, for any adopted value of e, this formula provides a second relation between f and ρ_m, and it is convenient to represent (3.1) in this way for a series of values of e since the resulting

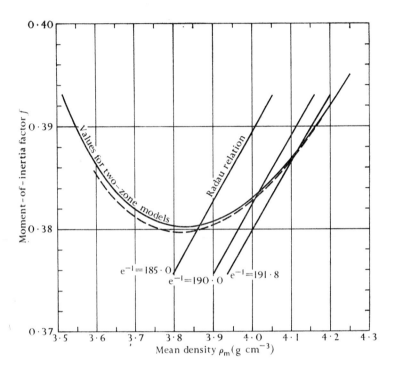

FIG. 3.1. Relation between moment-of-inertia factor and mean density for Mars. The solid curve is the graph of the moment-of-inertia factor f plotted against mean density ρ_m for two-zoned models of Mars; the almost straight-line curves show the Radau relation between f and ρ_m for three values of e^{-1} as indicated. The dashed curve shows how the relation between f and ρ_m for the models would be changed if outer layers departure from strict hydrostatic equilibrium increased the radius by 2·5 km.

graph of f against ρ_m happens to be practically a straight line. Fig. 3.1 shows the resulting curves: the continuous, roughly parabola-shaped, curve is based on the values of f and ρ_m of Table 3.3 (together with other computed values not tabulated there), while the straight lines represent the Radau formula for a few selected values of e.

Now it happens that the dynamical ellipticity e is the precise quantity that enters the expression for the rate of precession of the orbit of a satellite moving near an oblate planet. The value of e^{-1} derived by analysis of the motions of Phobos and Deimos is at present believed to be 191·8, but owing to errors of observation, systematic and otherwise, this could easily be in error by a few per cent. It is seen from Fig. 3.1 that this particular value would require a mean density of nearly 4·1 g cm^{-3}, which in turn would imply a radius of only about 3350 km. This seems inadmissibly low if the observations are to be trusted. On the other hand, if the mean density corresponding to $m_c = 0\cdot4$, or $r_s = 3398$ km, is adopted, the model would be consistent with $e^{-1} \simeq 188$, and this is probably within the range of uncertainty attaching to e. It may therefore be concluded that, in so far as present determinations of the mass, radius, and dynamical ellipticity of Mars are concerned, the internal structure of the planet is consistent with the general hypothesis that it is constituted of material similar to terrestrial material in the relevant elastic and physical properties. But it should certainly be emphasized that much more precise determinations of these quantities are highly desirable, especially for the radius and ellipticity, in order that more stringent tests can be applied to the theory of the structure of Mars. Also in need of investigation is the effect that the simplifying assumption of a sharp discontinuity between mantle material and outer-shell material may have on the calculated radius and other properties of these planetary models, for in the Earth this transition occupies a depth of some 300 km before the straight-line law $k = a + bp$ for the bulk modulus becomes satisfied with high accuracy.

For the very outermost part of Mars, the force of gravity is such that the ordinary hydrostatic pressure would not begin to exceed the strengths of the materials until a depth of about 10 km was reached; so there is the possibility of these outer layers receiving some measure of support from their intrinsic strengths in addition to that supplied by pressure. This would in general mean a slightly greater radius for the planet than would result on the basis of purely hydrostatic support. In addition, if lighter materials have come to the surface, as is believed to have occurred over at least parts of the Earth, and if weathering and erosion have produced finely divided material of lower bulk density at the surface, as seems to have happened on the moon, excess of radius over the calculated value by as much as 2 or 3 km could conceivably be maintained. If such difference lead to an increase of the overall radius by Δr, the resulting changes in the moment-of-inertia factor f and mean-density ρ_m are easily shown to be given by

$$\frac{\Delta f}{f} = \frac{2}{3}\frac{\Delta\rho_m}{\rho_m} = -2\frac{\Delta r}{r_s}. \qquad (3.2)$$

In Fig. 3.1, the dashed curve has been drawn on the assumption that $\Delta r = 2.5$ km, a value selected purely for illustration. It is seen that for a given value of e the changes produced are quite small, but nevertheless approach a lower value of the mean-density by amounts of the order of 0.01 g cm^{-3}. If the figure of 2.5 km should be an overestimate of the possible increase of radius, as is probably the case, the actual changes would, of course, be smaller, and much less than the uncertainties arising from imperfect knowledge of the mass, radius, and ellipticity. No serious modification of our earlier tentative conclusions for Mars would thus be implied by these possible surface effects.

MERCURY

Our knowledge of the planet Mercury is far weaker than for any other member of the terrestrial group. The mass is only about half that of Mars, which means about four or five lunar masses, and the planet moves so near the overwhelmingly powerful sun that its disturbing effects on other planets are extremely small. As the planet has no detectable satellite, its mass has perforce to be found from these minute planetary perturbations, and the probable errors of the various determinations so made are unavoidably large. Values found by different investigators range at least ± 10 per cent from their mean, as the results quoted in Table 3.4 show.

Table 3.4. Mass of Mercury from planetary perturbations

	Solar mass/Mercury × 10^6	Mass (10^{27})
Duncombe	5·88 (1 ± 0·077)	0·3379 ± 0·0260
Rabe	6·12 (1 ± 0·007)	0·3246 ± 0·0023
Brouwer	6·48 (1 ± 0·055)	0·3066 ± 0·0169

The maximum and minimum values implied simply by the probable errors of these results are seen to range from $0·3639 \times 10^{27}$ g down to $0·2897 \times 10^{27}$ g, but the actual mass could still lie outside this range. For numerical purposes a unique value must be selected, and if the mean of the three values is adopted this would give

$$M = 0·32304 \times 10^{27} \text{ g}, \tag{3.3}$$

but the probable error of this is rather more than 10 per cent, and it has to be remembered that the probable error itself means only an even chance of the true value falling within the stated limits. It is always possible that sooner or later a comet will chance to pass close to Mercury and undergo sufficiently large perturbations of its orbit to yield a greatly improved value of the mass, but it might be a matter of many centuries before so propitious an event occurred. A close fly-past by a spacecraft would be equally or even more effective, so here again we come to an important measurement that can be achieved within a reasonable time only by means of an interplanetary probe. Because of the proximity of Mercury to the sun as compared with the Earth (its mean distance is just under two-fifths that of the Earth from the sun), extremely high energy would be needed to approach Mercury directly from the Earth, but with available propulsion there is the possibility of utilizing an initial controlled encounter with Venus to deflect a spacecraft inwards on such a path that only small subsequent adjustments would be needed to cause it to pass as close to Mercury as may be wished.

` Not only is the mass of Mercury difficult to determine but so also is the radius, despite the fact that there can be at most only a minute amount of atmosphere. The planet is rarely to be seen at more than about 25° elongation from the sun and as a result always lies in a bright morning or evening sky near the horizon, while its angular size is seldom more than about 10 seconds of arc, which is only just over 1/200th the angular diameter of the moon. Russell, Dugan, and Stewart give for the actual radius

$$r_s = 2495 \pm 125 \text{ km}, \qquad (3.4)$$

which means a proportionate probable error of ± 5 per cent. Since the mean density is proportional to Mr_s^{-3}, the value for ρ_m implied by (3.3) and (3.4) would have a probable error of ± 25 per cent, so that

$$\rho_m = 4 \cdot 97 \pm 1 \cdot 24 \text{ g cm}^{-3}, \qquad (3.5)$$

and the range given by the probable error is from 3·73 to 6·21 g cm^{-3}. If a value as great as 4·97 g cm^{-3} were correct, it would mean that in spite of its much lower mass Mercury would have mean density far higher than Mars (3·92 g cm^{-3}) and almost equal to that of Venus (5·06 g cm^{-3} if $r_s = 6125$ km); the possibility of the planet having similar composition and structure to other members of the terrestrial group would then be remote. Attempts have already been made by some authors to devise explanations for such a difference, as for example that at such a short distance from the sun lighter constituents would be driven off, but it is more than doubtful whether causes should be sought for phenomena not definitely established observationally, for what becomes of the 'theory' should it eventually turn out that the phenomenon it explains is non-existent?

However, for the sake of completeness, the hypothesis that

Mercury *is* of similar composition to the other terrestrial planets can be utilized to calculate a series of models as for the other planets. Although the steady-state temperature would evidently be higher than for any other planet, radioactive energy would leak out more rapidly because of its small size, and as the central pressure would be only about two-thirds the value in Mars, it seems highly probable that Mercury is also solid throughout, and can therefore be represented by two-zone models in the same way as for Mars. Table 3.5 gives a selection of computed values for the various quantities.

Table 3.5. All-solid two-zone models of Mercury

m_0 $(10^{27}\,\text{g})$	ρ_0 (g cm^{-3})	p_0 $(10^{12}\,\text{dyn}\,\text{cm}^{-2})$	r_0 (km)	p_{int} $(10^{12}\,\text{dyn}\,\text{cm}^{-2})$	r_s (km)	f	ρ_m (g cm^{-3})
0·0+	4·296	0·1365	0	0·1365	2821	0·3955	3·437
0·05	4·328	0·1561	1408	0·1050	2796	0·3881	3·527
0·1	4·341	0·1639	1776	0·0829	2771	0·3836	3·625
0·15	4·349	0·1690	2036	0·0630	2745	0·3818	3·730
0·2	4·354	0·1723	2244	0·0441	2717	0·3826	3·846
0·25	4·357	0·1743	2421	0·0259	2687	0·3861	3·972
0·3	4·359	0·1752	2576	0·0081	2657	0·3926	4·110
0·32304	4·359	0·1753	2643	0·0	2643	0·3967	4·179

It is seen that the quantities follow similar trends, for increasing m_c, as do the results for Mars and all-solid models of the Earth and Venus, though now with somewhat smaller ranges of values because of the smaller total mass. The appropriate value of the interface pressure might well be much the same as for Mars, despite the smaller size, because of the proximity of the sun, which might mean a general temperature of some 400°, and so offset the more rapid escape of radioactively produced heat. The value of approximately $0.08 \times 10^{12}\,\text{dyn cm}^{-2}$, corresponding to the entry $m_c = 0.1$, suggests a radius as large as 2770 km, which exceeds the quoted observed value by a little more than twice the probable error and would place the mean density just at the lower limit of the probable-error range. On present evidence, therefore, if Mercury has similar composition to the Earth, the implication would be that the observed radius is underestimated and the mass overestimated. But there remains the alternative possibility of an essential difference in composition with different elastic properties and uncompressed densities. It will be a matter for the future to settle to what extent these different possibilities may hold.

THE MOON

Finally we may consider the application of the general idea of uniformity of composition to the moon. Here the question is of especial interest to hypotheses of the origin of the moon. If the Earth and moon have had an intimate common origin, closely similar compositions might be expected. On the other hand, if it emerged that their compositions differed in essential properties, then separate origins might have to be inferred, with capture of the moon by the Earth some time after their individual formations. However, even if similarity of compositions were established it would not rule out the possibility of subsequent capture, since it may well be the case that all the objects that formed in the inner reaches of the solar system had similar compositions anyway.

The mass of the moon in terms of that of the Earth is known with considerable accuracy from dynamical theory, the ratio being $1/81 \cdot 27$. The average radius of the moon is also reasonably well known, though there are certain special problems that arise connected with the determination of this, since the extreme proximity of the moon enables surface irregularities to be measured in much more minute detail than for any other extra-terrestrial object. According to Russell, Dugan, and Stewart, the observed average radius is given by

$$r_{\mathrm{s}} = 1737 \cdot 95 \text{ km.} \qquad (3.6)$$

The minimum possible central pressure in the moon then follows at once on the basis of uniform density as $0 \cdot 047 \times 10^{12}$ dyn cm^{-2}, and as this pressure is reached in the Earth at a depth of only about 150 km, less than half the thickness of the outer shell, it is plain that the moon can be regarded as a one-zone body, consisting of outer-shell material only, for the purposes of calculation. This means that only a single model will result, and the theoretical values of the principal quantities so found are as follows:

$$
\left.
\begin{aligned}
\rho_c &= 3\cdot427 \text{ g cm}^{-3} = 1\cdot025\ \rho_m \\
p_o &= 0\cdot0480 \times 10^{12} \text{ dyn cm}^{-2} \\
r_s &= 1736\cdot93 \text{ km} \\
f &= 0\cdot39823 \\
\rho_m &= 3\cdot350 \text{ g cm}^{-3}
\end{aligned}
\right\}
\qquad (3.7)
$$

These figures show at once that the main body of the moon departs only slightly from completely uniform density, for the central density and pressure exceed the values for a uniform moon by only a little more than 2 per cent. The calculated radius is just about 1 km *short* of the observed value, but a difference of this amount is scarcely surprising. Within the moon purely hydrostatic pressure would rise above the strength of the material at depths exceeding only about 20 km; material extending from the surface to a depth of this order may not therefore be distributed strictly in accordance with hydrostatic equilibrium, and any naturally arising departure from this will lead to some increase in the actual radius over the theoretical one. Moreover there is the possibility that the extreme outer layers of the moon consist of material with a peculiar history and of such a nature, through erosion and bombardment, as to be at much lower bulk-density than the standard uncompressed density. Taken together, these possibilities could well account for a difference of 1 km, and it is to be noted that the discrepancy is in the right direction to be so explicable; a value 1 km too large would have been a serious difficulty. It may be mentioned here that an entirely uncompressed moon of density $3\cdot298$ g cm^{-3} throughout would have a radius of $1746\cdot01$ km, or $9\cdot98$ km greater than the calculated value.

Where the moment of inertia is concerned, this can be determined with fair accuracy from observations together with the dynamical theory of the lunar rotation, and the value found is

$$f = 0.3971 \pm 0.0007. \qquad (3.8)$$

This also is seen to differ from the calculated value in the direction that would be expected, for according to (3.2) (p.101) an increase of radius by 1·02 km would change the value of f by -0.00047, reducing it to $f = 0.39776$, which agrees with the observed value at just about the upper limit of the probable-error range. Thus it may be concluded that the indications of theory are that there is no reason at present to regard the moon as having a composition essentially different from that of the Earth.

The central hypothesis of this chapter has been that the terrestrial planets and the moon are of similar compositions. It would appear that three main things have hitherto stood in the way of reducing the structure and evolution of these bodies to a consistent theory: first, the acceptance of the notion that the planets began as hot molten bodies; second, that the liquid core of the Earth consists mainly of iron; and third, but not least, that at all pressures the result of heating must be to produce expansion. It is the view maintained here that the masses of the Earth and Venus happen to be of just the necessary order for their central pressures to lead to liquefaction and contraction at the temperatures produced by radioactive heating, and that it is this that renders them such actively interesting and complex planets as compared with Mars and Mercury, which on the same physical basis are seen to have simpler physical structures without liquid cores, and accordingly to be less active. The theory, of course, is concerned with a realm of pressure much beyond anything yet reached in the laboratory as far as *static* pressures have been measured, and the importance to planetary physics of extending the range of experimentally obtainable pressures scarcely needs emphasis. But the relevance of temperature to these possible phase-changes must also be taken into account in any such experiments. It would also be valuable to have more knowledge of the factors involved in the 20°-discontinuity in the Earth, and of the physical conditions in those layers immediately below that level where the linear relation between k and p is not so closely satisfied as elsewhere.

On the astronomical side, obviously desirable would be more accurate knowledge of the mass, radius, and ellipticity of all the bodies concerned. When the landing of instruments on these objects takes place, refined measurements of any magnetic field present and the recording of any seismic disturbances will be of the highest interest, as also of course will be the detailed inspection of the solid surfaces themselves with the object of determining to what extent, if any, mountain-building has occurred. Regrettably this at present seems likely to prove most difficult for the one planet for which mountain ranges are predicted by the theory, namely Venus; but means may ultimately be devised for penetrating or avoiding the serious obstacle of its cloud cover.

How far the theory of planetary structure described and advocated here is correct will be a matter for the future to decide, but as it stands it not only offers a challenge to the time-honoured 'iron-core' theory, but it suggests experiments and observations that could lead to decisive conclusions. Meanwhile both hypotheses must be regarded as remaining in the field, and no final decision between them can or need be made until further crucial evidence becomes available from the discoveries of space-research.

4 The Nature of Comets

THE comets, often referred to as the 'second solar family', in some ways seem a kind of extra in the solar system. Study of them has not been intense in the past half-century for a number of reasons. On the theoretical side, even partially acceptable hypotheses as to their origin have only recently been forthcoming, and as comets found no recognizable place whatever in the cosmogonies of Jeans and others of his time, they pass almost without mention in their published works. On the observational side, the advent of the vast majority of comets, the so-called long-period comets, cannot be predicted, and professional astronomers can seldom prepare to observe what are usually very transient apparitions lasting perhaps a few months at most and occurring haphazardly through the years. However, with reviving interest in comets, it is encouraging that the great daylight comet of 1965, Ikeya–Seki, was for a few weeks subjected to intense observation both by professionals and amateurs. Regrettably, the predictable short-period comets, with the possible exception of Halley, are far too faint and unpromising objects, compared with the sun and other astrophysical bodies, to be made the subject of modern studies. As a result, even the work of discovering comets has been left largely to amateurs, though photographic discoveries are made more or less by chance from time to time by professional astronomers on plates usually taken for quite other purposes.

There are two main classes of comets. First, the short-period

group, about a hundred in number, almost all associated closely
with the orbit of Jupiter. As the name indicates, they have short
periods mainly of less than 10 years, and they nearly all move
in the direct sense in fairly eccentric orbits (average value about
0·5) and moderate inclinations to the plane of the Earth's orbit
(average about 15°). Halley's comet, with a period as long as
about 76 years, is an exception in that its motion is retrograde
compared with the planets. It has been demonstrated from
dynamical principles that these short-period comets represent
the remains of what were originally long-period comets that
chanced to pass sufficiently close to Jupiter for their paths to
be changed to short-period ellipses. The fact that in most cases
their paths still pass close to the orbit of Jupiter is a surviving
indication of this. But once so deflected, these comets must have
comparatively short lifetimes, astronomically speaking, and
probably no short-period comet can survive more than about
10 000 years. It will be described later how comets are thought
to consist of widely separated swarms of dust-particles, and the
kind of development that a short-period comet undergoes is one
of dispersal into a meteor-stream gradually extending more and
more round the orbit away from the comet, both ahead of it and
behind it in its motion. The end result consists solely of a meteor-
stream with no associated comet. Several meteor-streams are
known to be associated with particular comets, as for example
the Aquarids with Halley's comet, and the Taurids with comet
Encke. But there are also streams with no observed associated
comet. A specially interesting case is the Andromedids. These
move in the orbit of comet Biela, though this comet has not been
seen since 1852, when it appeared as two slowly separating
comets resulting from a division of the comet actually observed
at its previous 1846 return.

The far more numerous long-period comets, which may
number millions in all, are clearly to be regarded as the standard
type of object, though they exhibit great differences among
themselves. As their name suggests, their orbits are of long
periods, so long indeed that the value for any particular comet

is difficult to determine accurately. It is certain that the periods range from a few hundred years to many millions of years, but the longer the period the more difficult it is to determine it precisely, first because only a small part of the orbit close to the sun is covered by observations and there is always difficulty in measuring an accurate position for a comet, and secondly because of planetary dynamical action. Every year on average four or five of these objects come in to the inner reaches of the solar system and are newly discovered (though some may have been seen by prehistoric man at earlier returns); their periods average as long as 100 000 years. Of the order of 500 000 such comets, at least, are gravitationally bound to the sun at the present time and will eventually come sufficiently close to be observable from the Earth, and the number could be several times greater than this for various reasons. In the first place, a great many may approach the sun and depart again in a daylight sky and therefore escape detection. That this may happen frequently is shown by the fact that at least three times during solar eclipses unknown and uncharted comets have suddenly been seen close to the sun during the very few minutes of totality. Then again, there may be comets of large perihelion distance that remain always too far from the Earth for detection. Finally, so little is known or is likely to become known for a very long time of the distribution of numbers of comets against period, especially where very long periods are concerned, that the most probable total number can only be roughly estimated at between two and three million, though as extreme limits it is possible, but unlikely, that there are as few as half a million or as many as forty million. This wide range of theoretical estimates indicates how uncertain the matter is at present.

The general properties of comets are almost the direct antithesis of those of the planets. Comets move in highly elongated almost parabolic orbits, and the directions they come in from are distributed almost all over the sky, and while some move round the sun in the same sense as the planets, slightly more travel in the opposite direction. The overall extents of comets, as indicated by their observed comas, are in many cases far larger than the dimensions of planets, and a few are at times even comparable in volume with the sun. On the other hand, the masses of even the largest comets are negligible by planetary standards, and most comets probably have masses somewhere within the range 10^{15} g to 10^{21} g, so that ten thousand million average comets might be needed to equal the mass of the Earth. No comet has ever produced the smallest detectable gravitational influence on any other body, though instances are on record of comets passing quite close to planets and even to the satellites of Jupiter.

The consensus of opinion among the earlier comet workers, in an age when comets were a prime subject for observation by many professional astronomers, was that comets consist, when in their most quiescent state at great distance from the sun, of vast irregularly shaped swarms of widely spaced tiny particles of dust. The separations may be of the order of metres or more. The densities of the dust particles themselves, which may have a weak structure perhaps like snowflakes, would not be very different from unity, but the average overall density, allowing for their extensive volume, would probably be as low as 10^{-12} to 10^{-13} g cm^{-3}, though this may change systematically as the comet pursues its orbit. Lowell described a comet as 'a bagful of nothing'. This picture of their structure explains how it is that stars directly behind a comet can be seen undimmed right through the head; it also explains why certain comets have become invisible when they transited the disk of the sun, and how they can extend through such enormous volumes yet have such small total masses. No two comets behave in quite the same way, and individual comets themselves undergo such curious changes that identification of a comet with a recorded earlier one can be made only by means of the orbital elements. This must also include timing within the orbit, for there are numerous instances of groups of several comets, seven or more in one example, all moving in practically the same path, but passing through perihelion at quite different and unrelated times. About twenty such groups exist beyond doubt and there is some evidence for upwards of sixty. The number is likely to increase gradually as observations accumulate.

These strange properties of comets, although familiar to Newton three centuries ago, seem to have been largely lost sight of by some modern writers, and yet any theory to be regarded as satisfactory must show some possibility of explaining them. Leonardo da Vinci, more than 400 years ago, wrote: 'Why, this comet seems variable in shape, so that at one time it is round, at another long, at another divided into two or three parts, at another united, and sometimes invisible and sometimes becoming visible again.' This graphic statement describes many of the remarkable properties of comets, but perhaps more surprising still there occurs in addition for many comets a general *contraction* of the head or coma as the sun is approached. In the case of comet Encke, for example, which has now been seen at some fifty returns, the ratio of its largest volume to its smallest, as estimated from the observed linear dimensions, is sometimes as great as 100 000 to 1.

The apparent size of the head of a comet, which may appear irregular in outline, depends upon the means of observation, and is found to be different for telescopes of different power; but of the tendency to actual variability of size, and of the vast dimensions, there is no doubt. It is recorded of the appearance of Halley's comet at its return in 1910 that at 3 AU from the sun, when only faintly detectable, the coma was about 20 000 km across; at 2 AU it was about 300 000 km; at 0·6 AU it had shrunk again to about 200 000 km. After the comet had passed perihelion the observable coma appeared to expand again to reach about 400 000 km at 2 AU, and when very near the limit of detection at 4 AU it measured about 50 000 km across. At one extreme, comets less than 10 000 km across are extremely rare, though this may be due to the intrinsic faintness of such small comets. At the other extreme, several have measured more than 250 000 km across. The head of the great comet of the year 1811 measured 1 500 000 km ; while the faint nebulosity surrounding comet Halley in 1909 was nearly 1 000 000 km in dimensions; and that of comet Holmes in 1892 on occasion exceeded 2 000 000 km. Depending on its proximity to the Earth, a comet may accordingly exhibit quite large overall angular size; some have measured as much as a degree or more. Comet Lexell at one time was observed to be over 2° across, and the great sun-grazing comet 1882 II shortly after passing perihelion appeared to be surrounded by a sheath some 3° to 4° in length extending in the general orbital direction. Some of these changes result from instrumental effects, but some are undoubtedly real, for if we imagine a swarm of particles expanded to greater and greater radius it would eventually become invisible simply because the optical depth would decrease indefinitely as the size increased, and the swarm would become more and more transparent, especially at its outermost parts.

When a comet is first observed at great distance it has the appearance of a hazy silvery patch, perhaps a few minutes of arc in size, almost indistinguishable from the background sky. There is some resemblance to a nebula (which objects are very numerous in the heavens), but within a few hours a comet can be distinguished by its motion in the sky relative to the adjacent stars. The American astronomer, Barnard, who had remarkable power of eyesight, and was one of the most successful of comet-seekers, could *discover* comets that others could not then even see in the eyepiece! The light of a comet at the quiescent stage is purely reflected and scattered sunlight, and although some emission light may conceivably be present it is too faint to be detected spectroscopically. The total amount of reflected light, for a comet such as Halley, is such that only about 10^{-5} of its overall cross-sectional area can be producing any reflection, and this is a further indication of the widely spaced particle-structure of comets; if they were continuous bodies, even of a gaseous nature, of the observed dimensions, not only would they be far more luminous, but they would be of planetary mass, which is quite impossible. But as the comet approaches the perihelion portion of its orbit, the brightness increases far more strongly than can be accounted for merely by the decreasing distance from the sun. Not only does the amount of reflected light increase, but also a strong emission spectrum appears making a comparable contribution to the total luminosity. For most comets, the emission is associated with volatile radicals, but for those passing really close to the sun even the lines of heavy elements such as sodium and iron have been detected.

There is no clear-cut law of brightening applicable to all comets, but if the observed luminosity, L, at the Earth is assumed to follow a relation of the form

$$L = L_0/\Delta^2 r^n, \tag{4.1}$$

where Δ is the distance of the comet from the Earth, and r the distance from the sun, so that the whole intrinsic variation is associated with the changing distance from the sun, then it is found that r has an average value for all comets studied in this way of about 3·3, though in numerous instances it can be as high as 5, 6, or 7, or even more. There also occur seemingly quite irregular brightenings of a few comets, notably for Schwassmann–Wachmann 1 (1925 II), which besides being highly unusual among comets in moving in an almost circular orbit has often brightened up suddenly by several magnitudes —no less than about nine magnitudes on one occasion in 1946.

Again, as a comet approaches the sun there may appear to develop within the coma a small bright area or point of light, termed the *nucleus*. In some comets more than one nucleus may develop, but in others none is to be found at all. Estimates of the sizes of the nucleus in different comets range from a few thousand kilometres at one extreme down to immeasurably small points of light at the other. Despite the fact that a nucleus is not always present, many comet-workers have inferred that it must correspond to a small solid object deep within the coma, and they picture it as a small more-or-less spherical body composed largely of ice and meteoric dust particles. Attempts to measure optical sizes of nuclei produce widely different results, from thousands of kilometres in diameter down to values too small to measure, and even for the same comet the results vary according to both the means of observation and the observers themselves. But it is recognized by those favouring the small-snowball picture of the nucleus that any such object could at most be a few tens of kilometres in size, even if it is to contain the major portion of the mass of the comet, and above all if it is not to be directly detectable telescopically as an object of asteroidal dimensions. Observers of the Andromeda nebula report that the most luminous central region exhibits a bright point of light similar to the nucleus of a comet, but it appears that none have gone on to make the correspondingly valid or invalid inference that it consists of an icy snowball a few kilometres in diameter.

Even if the idea of a solid nucleus at the heart of a comet should prove correct, there is no reliable evidence for the hypo-thesis, and no such finite object could ever actually be seen short of sending a probe to the comet. Comet 1927 IV was fol-lowed telescopically to over 11 AU from the sun, which means beyond the orbit of Saturn, and at that distance a dusty snow-ball 10 or even 100 km in diameter (the black-body temperature would be only about 100°K or −170°C) could not possibly be observed in any telescope. Then again the great comet of 1729, which had a *perihelion* distance of over 4 AU and therefore came only slightly within the orbit of Jupiter, where the black-body temperature would still be less than 200°K (about −100°C), was visible to the unaided eye. It must have been a veritable giant among comets, perhaps intrinsically the largest on record, and there are no grounds for supposing that such a phenomenon could be produced from a tiny sphere of ice and dust warmed by solar heat. The general contraction of the coma as the comet approaches the sun, followed by re-expansion as it recedes, is also in conflict with what would be expected on the icy-con-glomerate model warmed by solar radiation. If anything, an *increase* in absolute size is what would result as the snowball developed an atmosphere, but quite apart from this it is difficult to see how an object 10^6–10^7 cm in size could come to produce an atmosphere for itself of dimensions 10^{10}–10^{11} cm. Besides not according with the properties of specific comets, the dusty-snowball picture of their structure provides not the slightest indication of the physical origin of the alleged nuclei. Nevertheless, so widespread appears to be the belief in a solid nucleus within a comet, that it must be one of the prime objectives of any spacecraft fired through a comet that it shall be capable of locating and measuring the properties of any nucleus present.

This uncertainty associated with the interpretation of the nucleus has important consequences where the accuracy of cometary orbits is concerned. In following a comet along its orbit, the position of the nucleus is usually regarded as defining the position of the centre of mass of the comet, and it is further supposed that the comet has a unique centre-of-mass moving in a Newtonian orbit. But if the nucleus is only an optical effect within an ill-defined luminous area produced by a vast disconnected swarm, none of these assumptions may be strictly valid, and a unique orbit for a comet may not even exist or be attainable. Certainly it is the case that cometary orbits are not known with anything like the precision of planetary orbits. The positions of the planets are known and can be predicted to better than a second of arc (1″), but it is doubtful if comets of large angular size could be correlated to an orbit within 10 seconds of arc, and where predictions are concerned they might be in error even by several minutes of arc. For the 1910 return of Halley's comet, the predicted time of perihelion-passage was in fact about three days too early, even though allowance had been made for all known actions.

The most sensational accompaniment of comets, when present, is the so-called *tail*. This can extend in observable form streaming out hundreds of millions of kilometres from the head, in the general direction away from the sun. In reality the tail material extends on indefinitely, but quite apart from the reduced incident solar radiation with increasing distance, the tail not only broadens out as it recedes from the comet, but is also speeded up by repulsive action from the sun, so that area for area its luminosity continuously diminishes along its length. But tails are by no means always present, and the general extent and intensity of tail-production vary enormously as the comet pursues its path, reaching an intense maximum at a time usually not long after perihelion, as also does the general total intrinsic brightness of the comet. Even when no distinct tail seems to be present, sufficiently long photographic exposure will nearly always reveal some tail-like emanation from the comet, and indeed for the majority of faint comets this may be as much evidence of the tail as can ever be detected. The light of the tail usually shows both continuous reflected solar light and emission; and undoubtedly because of the extremely low density of the tail there appears to be a higher proportion of ionized molecules than in the comet itself.

The geometrical form of the tail is seen only projected against the background sky, and the precise three-dimensional distribution within it cannot be determined. For highly luminous tails, the distribution is always clear-cut with a well-defined limiting edge on one side, but often with a quite ill-defined boundary on the other with the areal luminosity gradually diminishing. In some cases, depending on the position of the Earth in relation to the sun and comet, distinct curvature of this well-defined edge of the tail is present, but it must be remembered that what is being observed in the tail at any one time is the instantaneous arrangement of a whole volume of particles moving from the comet in hyperbolic orbits relative to the sun. Accordingly this curved edge can represent only the present position of the limiting line of particles and does not represent their actual paths. As all the particles of the tail move, so a new curved edge will appear in a slightly different position. In other words, the movement of the tail material will not in general be strictly along the tail itself. The lateral extent of the tail on its ill-defined edge is difficult to determine, but it may be extremely broad. Comet Chesaux showed six tails fanned out rather like a plume of feathers, and comet Borelly showed

PLATES 1–5

PLATE 1

a. Venus, photographed in blue light (200-inch telescope, Mt Palomar Observatory).

b. Jupiter in blue light, showing the Great Red Spot, the satellite Ganymede and its shadow (*above*) (200-inch telescope, Mt Palomar Observatory).

PLATE 2

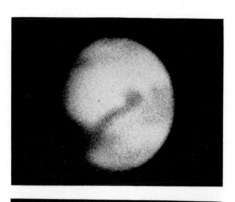

a. Mars in red light (100-inch telescope, Mt Wilson Observatory).

b. Saturn in blue light (200-inch telescope, Mt Palomar Observatory).

PLATE 3

a. Mars. The famous *Mariner IV* frame 11. About 25 000 square miles of Mars some 30° south of the equator, taken during the flypast on 14 July 1965. The space-craft was about 8000 miles away and the sun at an altitude of 43°. In addition to the obvious craters 10 to 20 miles across, there are clear signs of a much larger one about 100 miles in diameter. Evidently Mars (like the moon) has been subjected to meteoritic bombardment. (NASA photograph: Jet Propulsion Laboratory, California, USA.)

PLATE 3

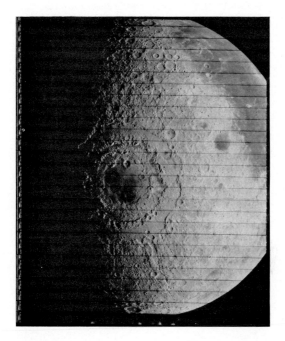

b. The lunar surface just at and beyond the western limb
visible from Earth. The Orientale Basin as photographed
from 1700 miles directly above it by *Orbiter IV* on 25 May
1967. The centre of this complex region is at about longitude 89°W and latitude 15°S, and from the Earth only the
eastern part is visible as the Corderilla mountains. The
outer circular scarp rises about 20 000 feet above the lunar
surface and measures some 600 miles in diameter. The whole
area may represent a comparatively recent formation, and
could well be the sole example of such a vast double-
rimmed feature still remaining exposed on the moon. (The
smallish dark irregular crater to the right is Grimaldi,
about 80 miles across and plainly visible from Earth.)

PLATE 4

a. Comet Arend–Roland (1957 III).

b. Comet Burnham (1959k), photographed on 27 April 1960.

PLATE 5

a. Comet Humason (1961e) (Yerkes Observatory).

b. Comet Ikeya–Seki, photographed on 29 October 1965 (Mt Wilson Observatory).

PLATES 6–8

PLATE 6

PLATE 7

Comparison of three natural australites with three artificially produced tektites. The actual diameters of the three artificial specimens (*top*) are 2·1 cm (*left*), 1·6 cm (*centre*), and 2·2 cm (*right*). Those of the three australite buttons are 2·5 cm (*left*), 2·4 cm (*centre*), and 2·3 cm (*right*). (NASA photograph: Ames Research Center, California, by courtesy of Dr. Dean R. Chapman.)

PLATE 8

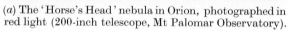

(*a*) The 'Horse's Head' nebula in Orion, photographed in red light (200-inch telescope, Mt Palomar Observatory).

PLATE 8

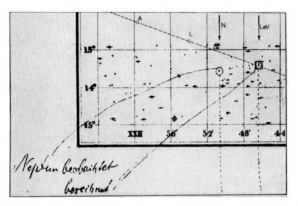

(*b*) Part of the original chart that enabled Galle in 1846 to find
Neptune (*N*) using Le Verrier's prediction (*LeV*). The
brightest star in this field (at RA $21^h 55^m 6^s$ and Dec
$-14°50'$) is ι Aquarii, mag. $4^m \cdot 3$. The handwritten entries
referring to Neptune's observed and calculated positions
(*bottom left*) are believed to have been made by Encke. In
this reproduction the ecliptic has been added, and the coordinates have been extended to Dec $-12°$, in order that
Adams's prediction (*A*) and the result of the Littlewood–
Lyttleton method (*L*) can be shown. (By courtesy of Dr. A.
Beer.)

photographically as many as nine tails. The possibility that the whole tail may occupy a wide angle, as seen from the comet, was strikingly illustrated by the comet Arend–Roland in 1957, when it happened that the Earth passed through the orbital plane of the comet. For a period of a few days, the comet looked as if there were a brilliant thin straight spike directly opposite the main tail, and the general line of the spike could be traced through into the central regions of the main tail. Evidently what was being seen was the result of a very thin flat distribution of dust and gas out to the opposite flank from that of the main tail, and this projected on to the background sky gave the appearance of a forward-projecting spike.

With the object of showing the tail at great distance, most photographs of comets considerably over-expose the head, with the result that it is difficult to decide just where the tail material emerges. But when the whole tail is very weak, it is of great interest to notice that its material is found to emerge only from the most central regions of the coma. For some comets the tail has seemed to detach itself almost completely from the head and then re-form in a matter of a few hours, as for example comet Morehouse of 1908. In others, the streaming motion of the tail sometimes becomes very irregular, as if the material had suddenly been swept aside by a gusty wind. This suggests that there may be minor clouds of gas moving independently in the solar system that are dense enough to be capable of interfering with what would otherwise be smoothly flowing tail material.

The record number of comets observed in any one year is twenty-two in 1947, made up of eight already found in 1946 and still remaining observable, eight new discoveries of long-period comets, four returns of earlier known short-period comets, one new short-period comet, and finally a suspected cometary object classified as doubtful. At the low record, no new comet was found at all in the year 1934, though seven already known ones were observable; and 1938 set a record of cometary inactivity in that only one was observed during the whole year, and even that was a previously known periodic

9—M.S.S.

comet. With long-period comets approaching the sun from practically all directions, the search for comets can be hopefully carried out in almost any area of the night sky. The most successful searchers belong almost entirely to the last century: it is believed that Pons discovered nearly forty, Brooks found as many as twenty, and Barnard nineteen. As something between 100 and 200 hours of searching is required on average to find a single comet and only a few hours work per night are possible, some idea of the devotion of these bygone astronomers to the task can be formed.

Despite the fact that these records belong to a bygone age, the actual rate of discovery is on the increase. In the decade ending 1950 over seventy discoveries were made. This was about twice the rate in 1841–50, which in turn was more than three times that of the corresponding decade a century earlier still. The average annual rate is now about six or seven, of which two or three are usually rediscoveries of known short-period comets, though an occasional new short-period comet may be found, or one missed at a number of previous returns rediscovered. As already noted, four or five represent entirely new discoveries of long-period comets.

The duration of visibility varies enormously from comet to comet, depending on the orbit and the relation of the Earth to it besides the intrinsic brightness. It may happen that two periods of observability occur as different parts of its path on opposite sides of perihelion bring the comet into the night sky. Most comets remain visible for a matter of months, but few for as long as a year, though 1936 I was observable for over two years, and 1927 IV as long as four years. At the other extreme comet Grigg in 1901 was visible only for twelve days; and a few very faint objects, suspected to be comets, have been observed only on a single occasion as far as is known. As very special cases, mentioned earlier, on three occasions a hitherto undetected comet has been observed during total eclipse of the sun, to become lost in the daylight sky immediately after totality had ceased. As to degree of visibility, the largest telescopes occasionally detect comets photographically at the very limit of their power, though almost all comets at some part of their orbit are completely invisible. Even intrinsically, however, there may be no lower limit to the size and brightness of comets. That the majority of comets have been detected with fairly small telescopes can probably be ascribed to the fact that most searchers for comets have made use of such relatively small instruments. Rather more than once in two years, a comet appears of sufficient brightness to become visible to the unaided eye somewhere on Earth; every five to ten years a really brilliant comet comes into view that is visible at night almost everywhere; and three or four comets per century are bright enough to be visible in the daylight sky. The most recent of these daylight comets was in the autumn of 1965: comet Ikeya-Seki, which proved to be yet another member of the celebrated group of sun-grazing comets. Previous daylight comets in the present century were in 1927 and 1910, the latter by great coincidence appearing shortly before comet Halley came closest to the sun. By all accounts, Halley's comet in 1910 was a disappointing spectacle compared with its 1835 appearance, and it seems probable that many of the enthusiastic descriptions of it recalled today relate in fact to observations of this 1910 daylight comet.

In addition to the phenomena associated with the increased luminosity and tail-development, several comets have been seen to undergo partition into two or more separate comets. This happened to Biela's comet in 1846, when the coma was observed to take on a pear-shaped form before an actual division took place. The two component-comets continued to travel in practically the same orbit, but by the end of two months had separated by about 300 000 km, and noticeable changes of brightness of both pieces occurred over a matter of days. At next return six years later, the separation along the orbit had increased to over 2 000 000 km, and although both comets were easily observable then neither has been detected since. However, the meteor-shower termed the Andromedids, sometimes the Bielids, after this comet, moves in a path closely similar to the original path of comet Biela. Until the disruption of Biela's comet was actually observed, the claim by Euphorus that the comet of the year 371 B.C. underwent disruption remained entirely discredited. But detachment of some portion of a comet has been observed on about ten occasions for different comets during the present century. The most remarkable case of all was the disruption of the daylight comet 1882 II, which, a few days after passing perihelion, appeared first to elongate into a brilliant streak some 180 000 km in length, within which were to be seen four or five star-like condensations, described at the time as resembling a luminous string of pearls. These condensations gradually separated from each other and were eventually identified as five separate large comets departing from the sun along practically identical orbits. The calculated orbital periods have been shown to range from about 600 to 950 years, and accordingly this group will return as four or five distinct comets at intervals of about a century.

Quite evidently, disruption of a comet in this way can result in the creation of a comet-group. This particular comet 1882 II when closest to the sun came within less than 500 000 km of the solar surface and passed through the perihelion side of its orbit in less than two hours. There can be little doubt that the observed elongation and disruption were brought about by the intense gravitational action of the sun on the different parts of the comet, and quite evidently disruption of this kind can result in the creation of a future group of comets. Indeed, 1882 II is itself an individual member of a group containing at least six other comets moving in almost identical orbits. But this group is unique in its extremely small perihelion distance q, and it is not permissible to suppose that all other comet groups, which have much larger values of q (see p. 136), have been formed through disruption of a larger comet at some remotely early perihelion passage.

A number of comets have mysteriously become increasingly diffuse and finally disappeared altogether as observable objects on their way *towards* the sun, when if anything it would be expected that they would become brighter and more readily detectable. This occurred for comet Ensor in 1906, for example, which early on was expected to become a conspicuous object, and again for comet Westphal in 1913. Perhaps related to such phenomena, there have been instances of very faint cometary bodies in orbits similar to those of known comets, but definitely displaced in time from the main comet. It may be conjectured that these represent detached portions only temporarily observable. There have also been numerous failures to recover short-period comets, and in present cometary lists there are about forty short-period comets that have been seen at only one apparition. An immediate explanation would be that such comets are short-lived anyway, but for at least some of them the failure may be due to an inadequately accurate orbit together with unfavourable location in relation to the Earth. With modern computing aids, accuracy of orbits is likely to be much improved, and already three short-period comets regarded as lost for several decades have been recovered. Periodic-comet Holmes of period of about 7 years, not seen since 1906, was found again in 1964 as an extremely faint object of magnitude 19; comet de Vico–Swift of period just over 6 years, last seen in 1895, was recovered in 1965; while comet Tempel–Tuttle of period 33 years, and last seen in 1866, evaded observation for just about a century. These rediscoveries suggest that application of modern computing techniques to cometary orbits could well result in the recovery of others, and until this has been thoroughly attempted it will not be possible to say what the real rate of loss to observation of short-period comets may be.

The total number of comets for which more or less detailed orbits have been computed approaches nearly 100 for short-period comets and nearly 500 for long-period comets. Since short-period comets almost certainly result from close encounters of long-period comets with Jupiter, the problem of the origin of comets is evidently concerned almost solely with accounting for the highly numerous long-period family. Up to about a decade ago, orbital elements had been computed for as many as 448 of these with periods in excess of 200 years. It is to be remembered that in computing an orbit, the aim is to obtain a Keplerian path about the sun—an ellipse, parabola, or hyperbola—that fits the observations as accurately as possible, and that these observations themselves are inevitably almost always restricted to an extremely small proportion of the entire path when the comet is fairly close to the sun and suitably situated in relation to the Earth. No attempt is made in any of the original calculations of orbital elements to allow for the detailed actions of the planets. The number of parameters— orbital elements—required to describe an ellipse or hyperbola and the position of the comet in it is six, but for various reasons an astronomer undertaking to determine an orbit from the available observations may in the first instance decide to regard it as a parabola anyway, for the period may be so long and the comet observed at so small a portion of its path that the adoption of all six elements would be an unjustifiable refinement.

Simple statistics of comets are of great interest, but it has always to be remembered that such data may be influenced strongly by selection effects. Thus, if the number of comets is plotted against perihelion distance q in the orbit, the resulting histogram shows a strong peak near $q = 1$ AU, and this is almost certainly an observational-selection factor. For smaller values of q, which would imply comets passing near the sun and therefore likely to become intrinsically much more luminous at perihelion, the number is fairly uniformly distributed, but for values of q above unity the number falls off rapidly, and there are few observed comets at all with q greater than 3 AU. As the intrinsic luminosity depends critically on distance from the sun, it is plain that with increasing q, other things being equal, a standard comet would appear far less luminous; and accordingly it cannot be inferred that the actual numbers with larger q-values fall off in the same way. It would require thorough analysis of the statistical data, supposing the existing amount adequate for the purpose, to decide the actual distribution with

q from the observed one. The largest known value of q is 5·52 AU for comet 1925 I, and there are four others with q in excess of 4 AU. The most remarkable perhaps was the comet of 1729, with $q = 4 \cdot 05$ AU; this was for a time visible to the naked eye, and may well therefore have been by far the largest comet ever observed.

Where the inclinations of cometary orbits to the plane of the ecliptic are concerned, the short-period comets must be considered separately from the long-period comets. For the former, practically all have moderately small inclinations, over nine-tenths being less than 35° and about half of them less than 10°, with an overall average of about 13°. Halley's comet is an exception in that, although its plane of motion is inclined at less than 18° to the general planetary plane, its direction of motion is retrograde so that technically the inclination has the supplementary value of 162°. For the long-period comets, the distribution conforms fairly closely to a purely random arrangement, apart from a few irregularities here and there, such as a slight deficiency in the range 100°–120° and an excess in 120°–150° over what would be expected on a purely random basis. It remains unknown whether these features require explanation. The nodes of the cometary orbits, that is the point where the path crosses through the ecliptic plane, appear to be distributed fairly uniformly all round the ecliptic, and suggest little or no departure from randomness.

The range of eccentricities of cometary orbits strikingly differentiates the family of short-period comets from that of the long-period ones. The short-period comets show eccentricities fairly uniformly distributed over the whole possible range. Until recently there were two so-called annual comets, Schwassmann–Wachmann I (1925 II) for which $e = 0 \cdot 136$ and comet Oterma with $e = 0 \cdot 144$, but the latter has recently undergone large perturbations changing its orbit considerably. These apart, half the total number have values of e lying between 0·3 and 0·7. In the range 0·7 to 0·9 are found about twenty comets and a similar number between 0·9 and 1. In this last set, only one exceeds 0·99, and in all only seven exceed 0·6, though all these latter have longish periods of the order of 100 years, as compared with the general run of short-period comets, which as has been mentioned, have average periods of only about seven years.

The foregoing values contrast strongly with the distribution of eccentricities of long-period comets, for which none is known with e less than 0·96. About a quarter of them all have e in the range 0·96 to 0·99 and the remaining three-quarters have values in excess of 0·99, with many crowded very close to unity. Although a value of e very close to 1 is characteristic of all known long-period comets, it is important to notice that for any small object of very long period this may be a condition for it to become observable at all. Thus if a comet has a period in excess of 10 000 years, as a great many have, its semi-major axis a must be at least 464 AU; and if in order to become observable its least distance $a(1 - e)$ must be less than 5 AU, which must certainly hold, then automatically e must exceed 0·99. So the possibility has to be recognized that there may exist numerous long-period comets that have remained undetectable through having large values of q, and if such do exist they may have values of e differing more from unity than is the case for observable long-period comets.

For the 448 computed long-period orbits, it has been found that 118 have a value of e less than 1, and 67 greater than 1, while for the remaining 263 cases a parabolic orbit, $e = 1$, was adopted at the outset. As a geometrical curve, the parabola extends to infinity, but in fact no comet has ever been found that has not originally come in from a finite distance. Despite the fact that all orbits are listed as if they were conic sections, it has to be recognized that no such path as an unchanging accurate ellipse, parabola, or hyperbola can exist for any body within the solar system because of the perturbing influence of the planets. However, if at any general instant of time all other attractions within the system except that of the sun are imagined to cease, a comet (regarded as a point-mass for the purpose) would thenceforward describe a unique ellipse (or parabola, or hyperbola) exactly coinciding at the instant concerned in both position and velocity with the actual path. Such an instantaneous orbit exists associated with every point of every path, however complicated, and it changes gradually as that path is described. For the planets, the changes are only small even over long intervals of time, but where the comets are concerned, precisely because e is very near to 1, only a small change in this element may be needed to pass from an elliptic instantaneous orbit (e less than 1) to a hyperbolic instantaneous orbit (e greater than 1), which means an infinite change in a.

The value $e = 1$ separates elliptic paths in which the object is always gravitationally bound to the sun from those paths in which it can escape from the solar system with energy to spare. When the attempt is made to fit the most accurate possible Keplerian path to the limited arc covered by observations, it is found sometimes that e comes out to a value slightly in excess of unity, so that the orbit of closest fit near perihelion, say, for a comet is in fact a hyperbola. These cases represent the so-called 'hyperbolic comets', of which, as mentioned above, nearly seventy have been found. But it cannot be inferred from this that these comets have come in directly from outside the solar system—from interstellar space, that is—because the action of the planets would be capable of accelerating the comet slightly on its way in and producing a temporarily hyperbolic osculating path. To investigate the matter, a far more extensive calculation is required for each such comet, making detailed allowance for the attractions of the planets, particularly the great outer planets, Jupiter, Saturn, Uranus, and Neptune. This has now been done for a large proportion of these hyperbolic comets, and in every such case the original path at great distance, out beyond the remotest planet, has been found to be truly elliptic when the accuracy of the calculations is good enough for there to be no uncertainty on that score. Associated with such calculations, based as they must be on unavoidably imperfect observations, there are necessarily numerical errors. For

But after allowance for planetary attractions, the *incoming* orbit at great distance was such that

$$\left.\begin{array}{l} 1/a = +0\cdot000856 \ (\mathrm{AU})^{-1} \\ e = 0\cdot998314 \end{array}\right\} \text{elliptic.} \qquad (4.4)$$

Then, carrying the calculations forward in time, the subsequent *outward* orbit at great distance was found to have

$$\left.\begin{array}{l} 1/a = +0\cdot000523 \ (\mathrm{AU})^{-1} \\ e = 0\cdot998973 \end{array}\right\} \text{elliptic.} \qquad (4.5)$$

These figures mean therefore that the comet came in from an original aphelion distance of just over 2300 AU, taking about 20 000 years to do so, and will make its way out again to a distance of about 3800 AU, taking nearly 42 000 years to reach its new aphelion. Thus, in this case, the osculating period at aphelion on the way in will have become more than doubled by the time the next aphelion is reached.

Large changes of period of this kind are characteristic of the effects of planetary perturbations on the orbits of long-period comets. In the example of the previous paragraph, elliptic inward motion was speeded up to a hyperbolic value temporarily near perihelion, and then the comet on its way out will slow down to elliptic motion again. But whereas all actual inward motions are elliptic, there is nothing intrinsic to the action of the planets to secure that the outward motion always reverts to elliptic; indeed, in a high proportion of examples now investigated the outward path remains *hyperbolic*, and such comets will therefore be lost for ever from the solar system. For example, for the long-period comet 1948 I, the following values have been found:

$$\begin{array}{ll} \text{Osculating orbit} & \left\{\begin{array}{l} 1/a = -0\cdot000411 \\ e = 1\cdot000307 \end{array}\right\} \text{hyperbolic.} \\[2ex] \text{near perihelion} & \\[1ex] \text{Inward path at} & \left\{\begin{array}{l} 1/a = +0\cdot000094 \\ e = 0\cdot999930 \end{array}\right\} \text{elliptic.} \qquad (4.6) \\[2ex] \text{great distance} & \\[1ex] \text{Outward path at} & \left\{\begin{array}{l} 1/a = -0\cdot000297 \\ e = 1\cdot000221 \end{array}\right\} \text{hyperbolic.} \\[1ex] \text{great distance} & \end{array}$$

The total number of cometary orbits studied in detail for the effect of planetary action now amounts to about sixty, and for more than one-third of these it is found that the ultimate outward path is hyperbolic; all these will therefore finally be lost to the solar system. The actual proportion of all long-period comets destined to escape at their next excursion from the sun is not so high as this, however, for the comets selected for these detailed studies have mainly been those showing hyperbolic motions, or very nearly so, in the observable part of the orbit, since the initial question stimulating such calculations was whether these comets had come in from outside the solar system.

It would in fact be expected that none of these comets has a truly hyperbolic entry path, for the eccentricities exceed unity by much too little for an extra-solar system motion to be likely. Relative to the stars in its neighbourhood, the sun has a speed of about 20 km s^{-1}, and so for any object entering under the purely dynamical attraction of the sun, the value of v ($= \mu^{\frac{1}{2}}a^{-\frac{1}{2}}$ for the hyperbolic path) must be of this order. On the other hand, the least distance from the sun is given by $q = a(e - 1)$, and so the eccentricity would exceed unity by qv^2/μ. With q of the order of the Earth's distance from the sun this quantity comes to about 0·5. To get a value of e such as 1·001, which is the kind of value exhibited by hyperbolic comets, would require with $q \sim 1$ AU an entry speed of less than 1 km s^{-1}; this could occur only very rarely if comets were simply wandering freely in interstellar space and occasionally encountered the sun.

This problem of the loss of long-period comets has been studied in great detail recently by purely theoretical means. Of the million or so such comets associated with the sun, probably less than one in a thousand has been observed, and as we have seen only for rather more than one in ten of these have the paths been worked out in detail. But because of inevitable uncertainties in the observations, it would not be possible to continue these detailed calculations for several revolutions and arrive at anything very meaningful. Unless the period of a comet could be calculated to an accuracy very much better than 1 year, the configuration of the planets at its next computed passage through the inner reaches of the solar system might bear little relation to the actual arrangement, and the subsequent pre-dicted orbit could be entirely erroneous. In fact, the periods may be in error by many thousands of years, and the nearer an actual comet comes to an escape path the less accurately is its period likely to be known. For these two reasons a theoretical study is the most that can be achieved, but the matter can be treated fairly simply by a statistical process. Because of the very long periods, the arrangements of the great planets at each successive return of the comet will be quite uncorrelated, and therefore the energy change that the comet undergoes at each approach to the sun may take some value randomly selected from all the possible energy changes that it could experience.

These energy changes and their distribution can be settled with considerable accuracy in several ways. A direct theoretical method is possible, making use of the Jacobi-integral of celestial mechanics for motion relative to two rotating centres of force—here the sun and Jupiter—and this gives results in complete accord with the energy changes computed numerically for individual actual comets. The general order of the changes can be regarded as changes in $1/a$ and measured in reciprocal astron-omical units. The following are typical values found from observed long-period comets:

$$
\left.
\begin{aligned}
&\Delta\ (1/a) \text{ from great distance to perihelion} \\
&\quad =\quad 630 \times 10^{-6}\ \mathrm{AU}^{-1}\ \text{(average for 39 comets)}, \\
&\Delta\ (1/a) \text{ from perihelion to next aphelion} \\
&\quad =\ -455 \times 10^{-6}\ \mathrm{AU}^{-1}\ \text{(average for 21 comets)}, \\
&\Delta\ (1/a) \text{ from one aphelion to the next} \\
&\quad =\quad 233 \times 10^{-6}\ \mathrm{AU}^{-1}\ \text{(average for 18 comets)}.
\end{aligned}
\right\} \quad (4.7)
$$

It is to be noticed that a binding-energy of the above order, given by $1/a = 250 \times 10^{-6}\ \mathrm{AU}^{-1}$, corresponds to a semi-major axis of 4000 AU and hence an orbital period of about 250 000 years. This is of the general order of the binding-energies of many long-period comets, and accordingly it is clear that they are subject to energy changes at each return that may be comparable with their own binding-energies. Study of the distribution of such changes, which depend on the cometary orbit and the position of disturbing planets relative to it, shows that they can be of either sign; so the comet can become more firmly bound to the sun, which automatically means a shorter period, or less firmly bound. If these random changes of binding-energy occurred at perfectly regular intervals or at randomly separated intervals, the evolution of a cometary orbit and the probable time before it escaped would be easy to follow statistically. But as the binding-energy, $1/a$, changes at each approach to the sun, so also does the period, which varies as $a^{3/2}$, and therefore the interval before the next change of energy occurs. The nearer the comet comes to escaping, the longer it is before the next return to the sun. It is this feature that complicates the problem and makes it mathematically difficult to determine the total time before a comet escapes.

But the difficulty can be surmounted by using a computer and simply following the development of the binding-energy for a sufficiently large number of comets in order to obtain average results. For example, if 250 comets are regarded as beginning with binding-energies such that $a = 1300$ AU, which means initial periods of about 47 000 years, then after half-a-million years about 50 per cent will have been expelled; after 5 million years about 95 per cent will have been lost in hyperbolic orbit; and after 50 million years over 98 per cent will have been lost; but complete removal of them all mathematically requires infinite time (Fig. 4.1, p. 133). These are average values for three different sets of 250 initial comets, but the several cases turn out to run an almost identical course—number lost against time—especially with increasing time. For weaker initial binding, the rate of loss is more rapid in the early stages, and conversely for stronger initial binding, as would be expected. But the influence of the planets eventually reduces any trace of the initial binding-energy, and the rate of loss eventually becomes almost independent of the initial values.

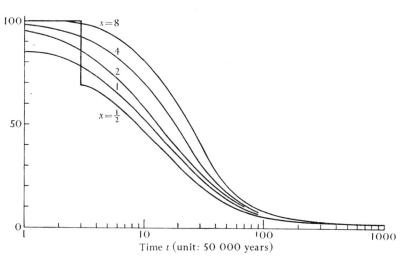

Fig. 4.1. Curves showing the rate of loss of long-period comets through dynamical action of the planets. The quantity x is the initial binding-energy of the comet in units such that the initial major axis is about $2600/x$ AU, and the five curves are for $x = \frac{1}{2}$, 1, 2, 4, and 8. The time-unit of the abscissa is 50 000 years with logarithmic scale.

In a time as short as 5×10^6 years ($t = 100$) more than nine-tenths of any group of comets would already have been expelled. The convergence of the several curves to the same form with increasing time results from the fact that the eventual binding-energies of surviving comets are determined mainly by the perturbing agency (the planets) and not by the initial values.

Analysis shows that, of a large number of comets, only one in about ten thousand could survive for a period as long as the age of the solar system. But the main way in which ejection is delayed for a comet is by its happening to get into an orbit of extremely long period, 10^8 years for example, so that it only seldom returns to the sun. By their nature, such comets are far less likely to be found, other things being equal, than comets of period say 10^5 years. For the latter will make 1000 returns to the sun for each single return made by the former. Observable comets have periods that average of order 50 000 to 100 000 years, so would represent only a small proportion of the remaining one in two thousand.

This result clearly has considerable bearing on the origin of comets. It has often been suggested that these curiously diffuse objects represent material that for some unexplained reason was not gathered into planets when these latter originated. But it seems certain for the reasons discussed that probably not more than 1 in 10 000 of such comets would have survived to this date. To leave nevertheless 10^6 comets would require an initial number of 10^9 to 10^{10}. This is not altogether impossible perhaps, but the probability of survival becomes smaller still through other processes of disruption. The dynamical ejection discussed above does not take any account of close approaches to planets and the resulting conversion of long-period comets to short-period ones, nor does it take account of the physical dissipative processes of tail-production and disruption into meteor-streams that may proceed quite rapidly for long-period comets. As is explained below, it is probable that a new crop of a few thousand comets is added to the solar system at regular intervals, of the order of a few million years, through a process of accretion from interstellar dust-clouds, a process that occurs quite independently of the number of comets already present. The consequence of this is that the total number of comets is probably gradually increasing at a rate approximately proportional to the cube root of the age of the system, so that for every 100 long-period comets that exist today there will be 125 when the system has doubled its age to 9000 million years.

Some remarkable statistical properties of the overall distribution of the known long-period comets emerge from a consideration of the directions in space from which they approach the sun. Since in a nearly parabolic ellipse the ratio of its axes, namely $(1 - e^2)^{1/2}$ is always extremely small, the path almost consists of two parallel lines of great extent turning back sharply at each end, and the direction of approach therefore coincides very closely with the direction of the aphelion point of the observed part of this orbit, which is diametrically opposite on the celestial sphere (centred at the sun) to the perihelion point. Fig. 4.2 shows the perihelion points of the 448

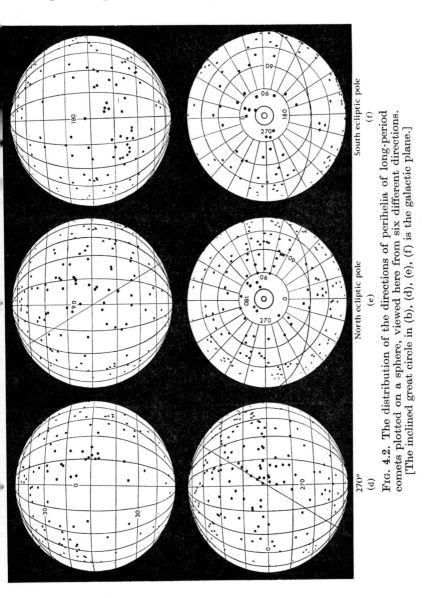

FIG. 4.2. The distribution of the directions of perihelia of long-period comets plotted on a sphere, viewed here from six different directions. [The inclined great circle in (b), (d), (e), (f) is the galactic plane.]

10—M.S.S.

comets, referred to above, for which reliable orbits are available. The diagrams show six views of a large sphere marked out in celestial latitude and longitude, λ and β, the marked parallels of longitude and latitude being at $15°$ intervals. Fig. 4.2(a) shows the sphere with the point $\lambda = 0$, $\beta = 0$ nearest to the camera and the ecliptic plane horizontal; Fig. 4.2(b, c, d) show views from longitudes $90°$, $180°$, and $270°$, respectively. These diagrams, which have been derived from photographs taken from a finite distance, do not in fact quite show the extreme polar regions of the sphere, but Figs. 4.2(e) and (f) show these areas photographed from directly above them. The line that appears in Figs. 4.2(b) and (d) as a great circle represents the position of the galactic plane.

Inspection of Fig. 4.2 immediately reveals a number of very striking features. First, the existence of large regions almost completely devoid of points, as for example that centred round the point $(350°, -20°)$ in Fig. 4.2(a), a zone first noted by Hoek almost a century ago; one centred at $(300°, -30°)$ in (d); and one at $(300°, 55°)$ in (e), to mention but a few. Secondly, there is the opposite feature of very compact groups containing several points, as for example the eight points near $(15°, 0°)$ in (a); four near $(105°, 12°)$ in (b); eight or ten near $(160°, -25°)$ in (c); seven near $(235°, 40°)$ in (d); and numerous other groups of varying degrees of compactness. When the other orbital elements of these groups are examined, they also are found to have closely similar values. On this criterion there appear to be as many as seventy-four such groups on the whole sphere containing from two to eight comets in each, but the criterion for deciding when two or three comets form a group is not really clear-cut unless the orbital elements agree extremely closely. To put any numerical rating on whether these features, of large empty areas and numerous close groups, are due to other than chance, the distribution must be subjected to some arbitrary statistical process. For example, when the sphere is divided up into seventy-two approximately equal areas covering its whole surface, there is found a preponderance of areas with many points and with very few points, and a deficiency of areas with the expected average of about six points. On a simple pigeon-hole argument, the probability of the distribution being due to pure chance turns out to be less than 1 in 1000, and any other tests to which the points are subjected yield a similar extremely small probability. This simply confirms the impressions obtained by mere visual inspection of Fig. 4.2.

If this marked tendency to grouping is real (and when account is taken also of the similarity of other orbital elements in a group there seems no possibility of its being due to pure chance) it has a surprising yet very simple implication. For where a group of, say, four or five comets is concerned, their observed approaches to the inner parts of the solar system have necessarily all taken place within the past 200 years or so; indeed, nearly three-quarters of all sightings are within the past century, and there is no reason to suppose that the present time is any way special as regards the number of comets associated with the sun. Accordingly, since the average periods are of the order of 50 000 to 100 000 years, this means that the whole group associated with just one of these directions of the celestial sphere may contain some thousands of individual members. Thus, the indications are that the long-period comets are to be thought of as consisting of a set of groups, of the order of a hundred or more in number, each containing a few thousand members, with all the members of each group coming in from the same general direction in space. But these directions themselves are widely distributed over the celestial sphere, though as will be explained later not in fact quite randomly. Within any one group, the positions of the perihelion points would be expected to diffuse gradually into a larger area on the celestial sphere as a result of planetary and stellar perturbations at each return, but the possible rate of this diffusion is a question that has not yet been investigated in detail analytically.

Certain other interesting results can be derived from numerical analysis of these perihelion points. For example, if all the points are given equal weight and any differences of distance from the sun ignored, so that they can be regarded as distributed on a sphere of unit radius, then for a random distribution of N points, it can be shown that the centre-of-mean-position would be most likely to be at an absolute distance less than $1/\sqrt{N}$ from the centre of the sphere; for 448 comets this in fact means less than about 0·043. Now it is a matter of simple arithmetic to find the coordinates of the centre-of-mass of the actual 448 points shown, and apart from insignificant errors of measurement the result is unique to the distribution. If rectangular coordinates centred at the sun are selected with OX through ($\lambda = 0°$, $\beta = 0°$), OY through ($\lambda = 90°$, $\beta = 0°$), and OZ at $\beta = +90°$, then with the sphere of radius unity, the centre-of-mass of the points is found to be

$$\bar{x} = -0·0069, \quad \bar{y} = -0·0446, \quad \bar{z} = 0·1316, \qquad (4.8)$$

and so the distance from the centre, $(\bar{x}^2 + \bar{y}^2 + \bar{z}^2)^{1/2}$, is 0·139. This is more than three times the expected value and would have only about a 1 per cent chance of occurring on a random basis.

As seen from the sun, this centre-of-mean-position will lie in a certain direction on the celestial sphere, and with the above values of $(\bar{x}, \bar{y}, \bar{z})$ this point is readily found to be

$$\lambda = 261°, \beta = 71°. \qquad (4.9)$$

The question immediately arises whether this corresponds to any known direction associated with the sun. In fact, it is only 20° away from the solar apex—the direction of motion of the sun relative to all stars in its neighbourhood—which has coordinates

$$\lambda = 271°, \beta = 53°. \qquad (4.10)$$

The chance that a point selected at random on a sphere should be within 20° of an already fixed point is less than 0·03. So this result suggests the interesting possibility that, despite the fact that the comets are gravitationally bound to the sun, their distribution is related to the solar motion in its local region of the galaxy. Whether this can be definitely concluded or not on this statistical evidence only, it is of interest that this position is the very opposite of what would be expected if the comets simply entered the solar system from interstellar space along purely dynamical paths. For the motion of the sun would then have the effect of making a greater number seem to come from the direction of the apex, so that the mean direction of approach would coincide with this direction, and therefore the mean perihelion direction, even allowing for some statistical fluctuation, should be in the general direction of the antipodal point on the celestial sphere.

The same discrepancy results for any hypothesis associating the comets with the neighbouring stars. For example, in default of any hypothesis for the origin of comets, it has often been proposed that a very large number—millions of millions of them—are permanently associated with the sun, moving in a vast spherical shell but remaining always at sufficiently great distances to be undetectable, and that from time to time passing stars deflect comets inwards from the shell by gravitational action. Encounters with passing stars would clearly be more numerous in the general direction of the solar apex, and accordingly the distribution of aphelion points would tend to have a maximum in that general direction, which again is the complete opposite of the observed distribution.

On the other hand, as will be explained in detail in Chapter 5 (see p. 145), the accretion process of the formation of comets from interstellar dust-clouds focuses this material into a stream in the axial line behind the sun in its motion through the cloud, a stream that may extend out radially from the sun many times the planetary distances. The comets then form as aggregates in this stream and fall in towards the sun from it. The aphelion points are thus in the opposite direction to the motion of the sun in the cloud; and so averaged over a number of encounters with clouds, the mean of the perihelion points would tend to coincide with the direction of the sun's way. The present statistical result may thus afford a crucial test that any theory of cometary origin must satisfy.

Yet another interesting property of the distribution of peri-helion points can also be readily established in answer to the question whether there exists any preferred plane near which the points have any tendency to congregate. Mere visual inspection of Fig. 4.2 suggests that this may be so, but the matter can readily be investigated numerically by adopting an arbitrary plane

$$Lx + My + Nz = 0 \qquad (4.11)$$

through the centre of the sphere, calculating the mean value of the square of the distances of all the points from it, and then choosing the ratios $L:M:N$ to give this its least value. Since the points are in fixed positions on the sphere, a definite result must emerge. It is found that the normal to this preferred plane is in the direction

$$L:M:N:1 \ = 0\cdot7674: \quad 0\cdot0151: \ -0\cdot6401: \ 1, \quad (4.12)$$

and again it is the case that this is close to a known astro-nomical direction. The pole of the galactic plane is in the direction given by

$$L':M':N':1 \ = 0\cdot8772: \ -0\cdot0536: \ -0\cdot4772: \ 1, \quad (4.13)$$

and the angle between these two directions is a mere $11\cdot5°$. The probability that this is a pure chance effect is about $0\cdot02$. The actual numerical value that emerges for the least mean-square distance is $0\cdot276$, whereas for a purely random distribu-tion on a sphere of unit radius it would be expected to be $0\cdot333$, with a standard deviation of $0\cdot014$. So the actual value found, $0\cdot276$, is in defect by more than four times the standard devia-tion, and the probability of this occurring on a purely random basis is considerably less than 1 per cent.

Thus here is a second strong suggestion from the observed distribution of these perihelion points that the comets are as-sociated with the galaxy and not with the general plane of the planets. The angle between these two planes, which is about $60°$, is sufficiently large, compared with $11\cdot5°$, for no possible confusion to be involved.

In statistical material of this kind, there is always the possibility of selection effects, though in the present case those that could readily be expected to be present are not such as would introduce any association with the galactic plane. For example, of the 448 comets utilized, as many as 259—about 58 per cent— have perihelion points north of the ecliptic, but this almost certainly results from the excess of observing stations in the northern hemisphere. If zones of equatorial latitude are taken, so that lack of perihelion points in the area near the south ecliptic pole is offset, the preferred plane not only remains near but is brought into closer coincidence with the galactic plane, and moreover the centre-of-mass of the distribution is shifted closer to the solar apex. Then again, if instead of using all the individual comets, each close group of three, four, five, six, or seven comets is replaced by a single point, the preferred plane is altered but little. The same holds true for the 127 comets of greatest brightness, selected as those observed for a period in excess of 4 months, except that for these the coincidence is still closer. This is also the case for comets of intermediate period, for which stellar and planetary perturbations are likely to be minimized; indeed, coincidence of the two planes is then reduced to less than 6°, suggesting that dynamical action tends to diffuse the positions of the aphelion points on the celestial sphere.

The light from a comet, besides varying enormously in total amount, may change very markedly in quality as the orbit is pursued. This is revealed by spectroscopic analysis whenever a comet is sufficiently bright to permit adequate dispersal into a spectrum. It is found that when a comet is quiescent, which is the normal state at great distance, the detectable light consists of a continuum crossed by the familiar solar Fraunhöfer lines and originating in the cometary spectrum through simple reflection, diffraction, and scattering, as would result from otherwise non-radiating small solid particles. The proportion of the total of such reflected and scattered light to the whole is usually found to be greatest in the bright central region of the coma and least in the tail, and the proportions vary strongly with the position in the orbit. Enormous increase in emission may take place as the comet approaches the sun, though it has to be remembered that at very great distances the total light may be too faint to permit any emission components to be disentangled successfully. Conversely, comets showing a very faint nucleus, as at times does Encke, may have no detectable continuous spectrum; but this could happen through great diffusion over a large range of wavelength of the total light.

The question of whether comets show any phase-effect is a difficult one to settle, not only because of the great changes of light that occur anyway, but because of the numerous physical factors likely to be concerned for a swarm of particles, such as could result from their individual sizes, shapes, and structures, and their degree of transparency. Some investigators have concluded that there is no dependence on phase-angle in the brightness of comets, but others claim to have discovered evidence of phase, though not of such a form as to suggest that the light is coming from a compact solid body. The associated question of polarization effects has also been investigated, notably for the bright comet Arend–Roland of 1957. This showed as high as 40 per cent polarization in the total light of the head, and the variation of polarization with changing phase-angle was considered to reinforce the photometric and spectrographic indications that this comet was largely composed of dust.

The emission spectrum of a comet usually consists of numerous bright bands of molecular substances composed principally of carbon, hydrogen, nitrogen, and oxygen in the form of compounds such as CH, CH^+, CH_2, CN, CO, C_2, C_3, NH, NH_2, OH, and OH^+. Occasionally comets approaching closer to the sun than average have shown lines of sodium, and sometimes those of magnesium, nickel, and even iron. The spectrum of the recent sun-grazing comet Ikeya–Seki, for instance, was at times rich in iron lines. The particular lines detected tend to show some correlation with distance from the sun; this supports the view that their formation is associated with the increasing strength of solar radiation, and the resulting black-body temperature, with decreasing distance. But this is not the only possibility, for the speed of the comet also increases as the radial distance decreases, and in a large dust-swarm this can have the effect of intensifying collisional effects between particles within the comet. It is of much interest in this connexion that recent studies have shown that when a molecule such as CO appears in both its un-ionized and ionized forms, the light associated with the latter is stronger deep within the coma than it is at the outer regions, indicating a greater proportion there of the ionized form. This is just the reverse of what would be expected if solar radiation were the source of energy of the ionization, and the phenomenon suggests a cause intrinsic to the comet. Collisions between some of its individual particles may take place at speeds sufficient not only to pulverize them into far finer dust, but to produce intense local heating on their surfaces, with the consequent release of volatile substances. The amounts of volatiles released would thus be in direct relation to the violence of the collisions.

Direct connexion between meteor-streams and short-period comets is now regarded as settled beyond question. In some instances, the annual recurrence of a meteor-shower indicates that the stream of particles responsible must be spread almost entirely round the cometary orbit, and also considerably dispersed laterally across the orbit, for the shower may last several days while the Earth is crossing through it. Such showers are usually weak. Indeed, several have been discovered that cannot be related to any known associated comet, but other much denser ones do not recur annually, and are presumably at an earlier stage of development of the spreading-away from the comet itself. All this is strongly indicative of the particle structure of comets. Moreover, the fact that there has been no instance of any object of meteoritic size reaching ground-level during any meteor-shower, however intense the shower may have been, further suggests that only small particles of insufficient mass to penetrate the high atmosphere are contained within comets.

5 The Origin of Comets

It is a curious feature of the problem of cometary origin that comparatively few theories have been advanced. For the planets themselves, multitudinous hypotheses at various levels of sophistication have been put forward, but for the comets, despite their impressiveness and the large amount of observational work devoted to them, little more than merely verbal suggestions have been made unsupported by mathematically expressed physical argument.

Obviously enough, any worth-while theory must rest upon proper appreciation of the phenomena concerned. Even the location of comets in space was not definitely known until the sixteenth century. It was possible for Aristotle to maintain that comets were a purely terrestrial atmospheric effect of the nature of a flame, and the fact that comets are brightest near the sun with their tails therefore stretching more or less upwards in the atmosphere seemed clear observational support for the idea. The great Ptolemy himself did not regard comets as celestial objects, but nevertheless there was diversity of opinion among the ancients on this fundamental question. It remained unsettled until the observations of Tycho Brahe, fairly soon to be fortified by the advent of the gravitational theory of the motions of comets, demonstrated the nature of their paths. But stable gravitational orbits reveal little or nothing of how the bodies concerned got into them. It has been shown in Chapter 1 how this difficulty presents itself for the planets, but there are huge numbers of binary stars in thoroughly stable elliptic orbits, and study of these discloses no immediate clue as to how such double-systems can have originated.

In the absence of any satisfactory theory of comets, numerous ideas have been put forward that have survived mainly through lack of knowledge of the properties, not only of the comets, but of the systems alleged to be responsible. The notion that comets have been expelled from the bodies of the great planets, Jupiter in particular, is an example. For this there is neither theoretical nor observational support; what little is known of the internal structure of planets is against the idea of ejection of a swarm of dust particles. As has been seen, the orbits of the long-period comets show no special relation to that of Jupiter, and nearly parabolic paths would be the exception and not the rule for such an origin. The association of most of the short-period comets with the orbit of Jupiter is of course undoubted, and their origin through occasional repeated deflections of long-period comets that chance to pass near Jupiter has long since been convincingly demonstrated on dynamical principles; but this in no way explains the physical formation and the original source of comets, and as we have seen it is this question of the origin of the long-period comets that represents the essence of the problem.

The allied notion of an origin in solar prominences can be passed over without discussion, but the idea has also been proposed that comets originate from asteroids in passing sufficiently close to a planet for differential gravitational action to bring about their disruption. The hypothesis at least has the initial merit of starting from safe ground, for the existence of large numbers of asteroids is a settled matter, but many comets possess orbits that do not pass anywhere near any planetary orbit, and of those that do few if any come so near as the Roche-limit for disruption would require. Then again, even if disruption occurred, cohesive forces would prevent really small pieces being formed, and it appears that an asteroid 10 km in radius, which would have roughly cometary mass, could graze the surface of Jupiter without undergoing disruption.

Yet another suggestion for an origin of comets within the solar system, one that has been put forward frequently from time immemorial, is that they represent the uncollected debris left over after the original formation of the planets. In quality of argument this is *ignotum per ignotius* in a big way, though this alone would not of course dispose of the suggestion. However, the idea is opposed to some of the most elementary properties of the comets. The common direction of motion of the planets and thousands of asteroids round the sun strongly suggests that the same feature characterized all elements of their material at an earlier stage, so that even material left over would be expected to show some sign of such common motion, whereas the distribution of the cometary orbits shows no such feature at all. Quite as many comets proceed retrogradely as forward; indeed, as we have seen, the orbits show no correlation with the invariable plane of the planetary system. Then again, the physical rate of dissipation of comets, by tail-production and dispersal into meteor-streams, appears to be far too rapid for them to have existed since the time of formation of the solar system. In the whole age of this system, a comet with average period 100 000 years would make 4.5×10^4 returns to the sun, and if at each one of these it lost only 1/1000th of its mass, through tail-formation and meteor-stream production, the initial mass would have been more than 10^{19} times as great as the present mass—which at a minimum means several times the mass of the sun! Even among comets of short-period, none has made more than twenty or thirty observed returns, with the exception of Encke, and the loss of mass through these physical causes could not be expected to have produced any systematic detectable effects over the periods of time involved, especially by comparison with the irregular effects inevitably introduced by the differing circumstances of separate returns and the imperfection of early records of returns.

The dynamical results referred to in the last chapter also tell strongly against origination simultaneously with the planets, in that only about one in 10^3 or 10^4 comets could survive hyperbolic ejection through the action of the great planets. Then again, although the problem of the rate of diffusion of perihelion and aphelion points on the celestial sphere as a result of planetary and stellar perturbations has not yet been worked out in full numerical detail, it seems highly probable that in a time of the order of several aeons practically all trace of grouping would be removed and a random distribution over the celestial sphere produced, or at any rate one showing correlation with the planetary plane rather than with the galactic plane as is the actual situation.

But where the distribution of binding-energies of comets is concerned, that is of $1/a$, these must even now show signs of planetary influence, though the ultimate result of such action is complete removal of all comets from the solar system. In undergoing changes of $1/a$ comparable with itself, and hence of a, the comets show yet another property exactly opposite to that of the planets. The reason for this lies in the fact that for comets e is very close to unity, whereas for planets it is very close to zero. To see in a general way how this comes about, consider the angular momentum per unit mass, h, of an object in a Keplerian orbit about the sun. This is given by

$$h = \sqrt{(\mu a |1 - e^2|)}, \qquad (5.1)$$

where μ is the mass of the sun in suitable units. In a partial sweep round the sun of a small planet like Mars, say, or a comet describing the inner part of its orbit, the change in h produced by the action of the great planets will be of much the same order, so that for both types of motion

$$2\mu^{-\frac{1}{2}}\delta h = a^{-\frac{1}{2}}|1 - e^2|^{\frac{1}{2}} \delta a \pm a^{\frac{1}{2}}|1 - e^2|^{-\frac{1}{2}} \delta(e^2). \qquad (5.2)$$

Now in an equation of this kind, the various terms for any actual motion are found to have sizes of much the same order. The order of magnitude of the changes δa and $\delta(e^2)$ can thus be found, for given δh, by taking each term separately as of the massive centre of force raises mathematical difficulties that remain of an almost insuperable character even when the problem is idealized by assuming a perfectly uniform cloud of negligible self-attraction streaming past the star with initially common uniform rectilinear motion at all its parts. From general considerations, it is plain that the attraction of the star will draw in the material sideways, and tend to focus it into the axial line behind itself, but the resulting compression will heat the gas, changing some of the energy of streaming and gravitational infall into random thermal motions. Were it not for this heating, there would be nothing ideally to prevent the material focusing into a line, the transverse components of velocity being lost as it met material converging from the opposite direction, and leaving only the resultant radial velocity. This is what would happen to begin with, and if the residual radial velocity were insufficient for escape from the star, the material would be captured. Even this process of line-accretion, as it has been termed, presents great difficulties of analysis, and it has been successfully solved only for steady-state motion.

As far as the initial hypotheses of the process are concerned, they are well founded in observation. The existence of obscuring dust in interstellar space seems first to have been recognized by Barnard (himself a great comet observer also), and it is now accepted without question that the galaxy abounds with irregular dust-clouds, of dimensions measured in parsecs, but of extremely irregular shapes and sizes. In overall volumetric extent they occupy some 10 per cent of the galaxy and, like the highly luminous stars that probably produce them when they shower off material in supernova explosions, they show concentration towards the galactic plane. Much of our own galaxy is obscured optically by these very clouds, but external galaxies show the overall distribution well, and in those that happen to lie edge-on to us the concentration to the equatorial plane of the system is often extremely striking. The dust-clouds are usually found to be in association with denser gaseous clouds, which are believed to consist mainly of hydrogen, and to which the dust contributes only of the order of 1 per cent of the mass. The densities of such gas-clouds are not known with high accuracy, but probably most of them lie in the range of 10^{-23} g cm^{-3} to 10^{-21} g cm^{-3}, though from their irregular appearance it can scarcely be expected that the density within them approximates to uniformity. Thus the density of the dust-clouds can be expected to be in the range of about 10^{-25}–10^{-23} g cm^{-3}, and again the indications are that the density distribution is far from uniform.

It is certain for a number of reasons that most of the particles going to form the dust-clouds are extremely minute. The clouds can therefore be thought of almost as mere smoke in space. The absence of diffraction and polarization effects, together with the simple reflection of starlight from the dust-clouds, suggest that most of the particles have sizes less than about 10^{-3} cm and greater than 10^{-5} cm. The mere fact that some of the clouds are completely opaque also indicates that most of the particles must be extremely small, for otherwise the overall density would need to be impossibly high. Thus, if all the particles had radius of the order of 10^{-1} cm, a density of 10^{-19} g cm^{-3} would be needed as a minimum for opaqueness to result from a depth of the order of 1 parsec. On the other hand, the distribution of sizes could be such that, while most of the particles were small, much of the mass was accounted for by the larger particles. Thus, a particle of radius 10^{-1} cm will have mass 10^6 times that of a particle of radius 10^{-3} cm, and there could consequently be 10^5 times as many of the smaller particles while the larger nevertheless contributed nine-tenths of the total mass.

Let us now consider what would be involved if the sun, in its orbit round the galaxy, encountered such a cloud. All the objects, both stars and clouds, in the general neighbourhood of the sun have a general orbital velocity of order 250 km s^{-1} transverse to the direction of the centre of the galaxy and more or less parallel to the galactic plane, but just as planets do not move round the sun in exactly the same plane nor with

11—M.S.S.

precisely the circular speeds appropriate to their distances, so slight differences of velocity exist among all these objects in the galaxy. This is why the sun has some velocity relative to any other star in its neighbourhood, but of much smaller value than 250 km s^{-1}; and in just the same way the sun would be expected to have a velocity relative to any dust-cloud. The direction of this relative velocity will not, of course, bear much relation to the direction of general circular motion in the galaxy, since it represents the difference between two almost parallel velocity-vectors each of magnitude some 250 km s^{-1}. But as the dust-clouds are likely to be moving in or close to the galactic plane, the distribution of relative velocity-vectors would tend in the long run to be symmetrical about this plane. It is also to be noted that as the sun travels within the cloud the relative velocity may change somewhat, for the cloud itself may not be in a state of pure translation but may have some internal motions, perhaps of the order of 1 km s^{-1} for widely separated parts. The accretion process may thus vary somewhat in its range and effectiveness during a single passage through a cloud.

Let us consider now, with Fig. 5.1 in mind, what will happen when the sun first enters such a cloud. It is easier to think of the sun at rest and the particles of the cloud moving by, all of them starting with equal parallel velocities when at great distance. In the undisturbed cloud, the dust particles will be widely separated. For instance, if to be specific $r = 10^{-4}$ cm is taken as the radius of a particle, and $\rho = 10^{-24}$ g cm^{-3} as the density in the cloud, then the separations would be of the order of 0·2 km. A particle of this size travelling through the undisturbed cloud would have very low probability of undergoing collisions; for a collision to be certain, the particle would in fact have to travel a distance measured in parsecs. The relevant parts of the orbits of the particles, while being deflected by the sun, can therefore be regarded as undisturbed hyperbolic paths. But when the material converges to the neighbourhood of the axial line behind the sun, the probability of a particle undergoing a

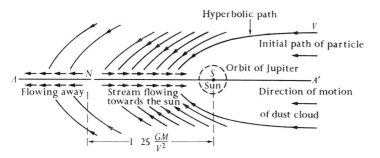

FIG. 5.1. Diagram (not to scale) illustrating the mechanism of accretion of interstellar dust by the sun.

The figure represents a section through the accretion axis AA'. The dust particles move round the sun in hyperbolic orbits and converge to the axis, where a stream of material is built up. The stream flows towards the sun within a distance $\sim 1{\cdot}25\ GM/V^2$, and away from the sun at greater distances.

The distance SN would actually be of the order of 100 times the radius of the orbit of Jupiter (whose plane would not necessarily or usually coincide with any plane through the axis). Also, the width of the stream would be only a small fraction of the solar radius, and is unavoidably much exaggerated here.

collision is enormously increased. Ideally, if we follow the motion as if it were that of a fluid, material lying on a surface when at great distance comes to occupy a line at the accretion axis (SN of Fig. 5.1), and an infinite increase of density would occur. In practice this is avoided as a result of a number of factors: first, the dust is concentrated in randomly distributed particles that do not necessarily arrive at the axis at precisely simultaneous times; and secondly, slight internal motions occur within the cloud, and planetary perturbations of the paths of the particles past the sun also take place. These latter are of the order of a few kilometres at most; and so instead of the paths of the particles all precisely crossing the axial line SN, they can be visualized as crossing through a cylinder of radius, say, 10^6 cm surrounding this axis (Fig. 5.2, p. 154). Looked at from a direction along the accretion axis, the paths would be seen as a whole set of lines converging in the general direction of the axis but missing it randomly, though all would pass within a small distance say, s, from it.

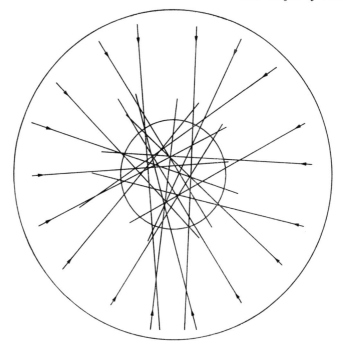

FIG. 5.2. Diagram (not to scale) showing the paths of particles converging towards the accretion axis but not quite intersecting it, as seen from great distance along the axis. The diagram cannot be drawn to scale: the radius of the outer cylinder will be of order GM/V^2 (the capture radius), while the radius of the inner circle will be determined by planetary perturbations and other irregularities of motion. It will be less than 10^{-9} of the former.

The probability of a particle undergoing a collision in crossing through this inner cylinder proves to be independent of its radius, and accordingly the accretion process remains unaffected by slight departures from the ideal undisturbed motion.

It turns out that the probability of collisions is independent of this deviation s when this is small compared with the general scale of the motion, which as will be seen is in excess of 10^{15} cm. For suppose we consider what happens within a cylinder of radius s and length b surrounding the axis. It is easily shown that the amount of material entering (and leaving) this cylinder in unit time is

$$2\pi\rho GMV^{-1}b, \tag{5.5}$$

where ρ is the density of the cloud at great distance, GM the strength of the sun's attraction, and V the velocity of the cloud at great distance relative to the sun (Fig. 5.1). The time that an individual particle would take to cross the cylinder (if it did not undergo a collision) would be of the order of $2s/V$, and for this time the particle would be contributing to the average volume-density within the cylinder. Hence in these initial stages the total mass in the cylinder at any instant would be

$$4\pi\rho GM V^{-2}bs. \tag{5.6}$$

The volume-density is this expression divided by $\pi s^2 b$, or

$$4\rho GM V^{-2}s^{-1}. \tag{5.7}$$

The particle-density within the cylinder is therefore

$$4nGM V^{-2}s^{-1}, \tag{5.8}$$

where n is the particle-density in the undisturbed cloud at great distance. If S is the effective collision area of one particle with another ($4\pi r^2$ for spherical particles), a particle in crossing the cylinder will sweep out a volume of the order of $2sS$. The number of particles it would encounter would be

$$(4nGM V^{-2}s^{-1})2sS = 8nGM V^{-2}S, \tag{5.9}$$

which is independent of s.

It is easy to see that this should be so, for the probability of collisions within the cylinder is a purely geometrical quantity independent of time. If the inner circle of Fig. 5.2 is imagined to be enlarged, or reduced, without the size of the particle being changed of course, the number of intersections that the selected path makes with others is unaltered, even though they become more widely or more closely spaced. It is as if a motorist decided to drive over, say, ten cross-roads without stopping: the probability of a collision at some one of them is the same whether they are spaced a hundred yards apart or a mile, or any other distance sufficiently smaller than his whole journey to encompass the crossings.

To arrive at an actual estimate of the probability, some definite numerical values must be adopted and some assumption made as to the shapes of the particles. If they were spherical of density σ g cm^{-3} and radius r cm, then $n = 3\rho/4\pi r^3 \sigma$, $S = 4\pi r^2$, and the expression (5.9) becomes

$$24\rho GM V^{-2}r^{-1}\sigma^{-1}. \tag{5.10}$$

With $V = 10^5 v$ cm s^{-1}, so that v is the speed in km s^{-1}, this becomes

$$3\cdot2 \times 10^{17}\rho v^{-2}r^{-1}\sigma^{-1}. \tag{5.11}$$

As interstellar particles may be of irregular shapes and weak structure, perhaps resembling snowflakes, the value to be assigned to σ is uncertain, except that it is likely to be of the order of unity. If for numerical illustration we adopt

$$\rho = 10^{-24}\,\text{g cm}^{-3}, \quad v = 5\,\text{km s}^{-1}, \quad r = 10^{-3}\,\text{cm}, \quad \sigma = 2\,\text{g cm}^{-3},$$

this gives an initial probability of collision for every particle of the order of 10^{-5}. Even if this probability were 10^{-10} or less, the number of particles arriving is so great that some collisions would begin to occur from the outset with effectively complete certainty. With the numerical values adopted above, the number of particles arriving per centimetre of length of the accretion axis is about 10^{-2} per sec. Thus in 10^6 seconds there would be of order 10^4 particles arriving, and it would be practically certain that a collision would occur. And this holds for every single centimetre of the axis.

But it is to be noticed that once collisions have begun to occur, the transverse velocities of the particles concerned will be reduced, and there will be an immediate tendency for material to be left in the collision zone surrounding the axis. The presence of this material will obviously increase the probability of further collisions, and in fact it will do so in a rapid exponential way. To see that this is so, if $p(t)$ is the probability of an incoming particle undergoing collision at time t, and $n(t)$ the total number of collisions (in unit length of the axis, say) occurring up to time t, then $p(t)$ will at least be proportional to $n(t)$, and since the number of collisions increases at a rate proportional to the probability of their occurring

$$\frac{dn}{dt} \propto p(t) \propto n(t) \tag{5.12}$$

with positive factors of proportionality. Accordingly

$$p(t) = p_0 e^{kt} \text{ where } k > 0, \tag{5.13}$$

and so long as p_0 is not strictly zero, the probability of further collisions begins to build up exponentially. Of course, since p cannot exceed unity, this simple analysis will not apply indefinitely, but the aim here is to show that despite the seemingly small initial probability of collisions, of order perhaps 10^{-5} per particle, the probability increases very rapidly and will soon attain to almost complete certainty.

When it is remembered that there is likely to be present in addition some hundred times as much gaseous matter, these probabilities are much further enhanced. Such gas will not so closely surround the accretion axis as would dust alone, but the extent to which it will assist in arresting the transverse motion of the particles will clearly be more or less independent of this. Also, once the accretion stream has begun to build up, a high concentration of matter is produced round the axis, flowing along it at just such a rate as is compensated by the further material coming in from the sides. The mass of material at any part of the stream, its line-density, in the steady-state motion can be calculated, and is fixed once ρ and V are fixed, but it seems likely that the transverse width of the stream will increase until it is such that particles are still almost certainly trapped by it.

As will be explained below, comets are formed as gravitational accumulations within this stream, and accordingly the general size of a comet at this stage would be expected to be of the order of the width of the stream. An order-of-magnitude estimate of this can be made in the following way. The line-density in the stream near the neutral point, which occurs at a distance $\alpha GM/V^2$ from the sun (where $1 < \alpha < 2$), is given by $2\pi\alpha^2\rho G^2 M^2/V^4$. If the radial width of stream perpendicular to the axis is w cm, the mass-density within it would be given by $2\alpha^2\rho G^2 M^2 V^{-4}w^{-2}$, and hence, for spherical particles of radius r and density σ, the number density N would accordingly be

$$3\alpha^2\rho G^2 M^2/2\pi V^4 w^2 r^3\sigma. \tag{5.14}$$

The probability of collision of a particle in describing a chord of (average) length $4w/\pi$ across the cylinder will be $(4w/\pi)N.4\pi r^2$, and this will be approaching certainty if

$$w \sim 24\alpha^2\rho G^2 M^2/\pi V^4 r\sigma. \tag{5.15}$$

Inserting $\rho = 10^{-24}$ g cm^{-3}, $V = 5$ km s^{-1}, $r = 10^{-4}$ cm, and with σ expressed in g cm^{-3}, this leads to $w \sim 2 \times 10^{10}\alpha^2\sigma^{-1}$ cm, which as we have seen, for α and σ of order unity, is of the same general order as the linear dimensions of comets as indicated by their observed comas.

Returning now to the accretion process itself, the equations governing steady-state motion are simple to set up when the stream is sufficiently narrow to be representable by motion in a line. If the constant amount of material entering from the sides is denoted by A, so that

$$A = 2\pi\rho GMV^{-1}, \tag{5.16}$$

and if u is the outward velocity at distance r from the sun, and m the line-density of material in the stream, then conservation of mass leads to the equation

$$\frac{d}{dr}(mu) = A, \tag{5.17}$$

while conservation of momentum leads to

$$\frac{d}{dr}(mu^2) = AV - GMmr^{-2}. \tag{5.18}$$

Since A is constant, (5.17) integrates at once to

$$mu = A(r - r_0), \tag{5.19}$$

where r_0 is some constant length associated with the motion. Its determination is of special importance, since evidently the stream is flowing towards the sun for r less than r_0, and conversely; $r = r_0$ therefore represents a neutral point (the point N of Fig. 5.1, p. 153) in the motion. It can be shown that r_0 is always of order GM/V^2, which is the characteristic length associated with the accretion process, and it can further be shown that for stability of the motion

$$GMV^{-2} < r_0 < 2GMV^{-2}. \qquad (5.20)$$

It is convenient in the mathematical analysis of the problem to introduce dimensionless variables x, y, and z replacing r, u, and m, respectively, and defined as follows:

$$\left. \begin{array}{ll} r = GMV^{-2}x, & r_0 = GMV^{-2}\alpha; \\[2mm] u = Vy, & m = 2\pi\rho G^2 M^2 V^{-4}z, \end{array} \right\} \qquad (5.21)$$

so that α replaces r_0. The numbers x, y, and z are then found to satisfy the following equations, equivalent to (5.18) and (5.19)

$$\frac{dy}{dx} = \frac{1-y}{x-\alpha} - \frac{1}{x^2 y}, \qquad (5.22)$$

and

$$yz = x - \alpha. \qquad (5.23)$$

These selected units are such that the values of x, y, and z are all of order unity at the region of the solution of (5.22) and (5.23) near the neutral point. For instance, at the neutral point itself

$$x = \alpha, \quad y = 0, \quad z = \alpha^2. \qquad (5.24)$$

Equation (5.22) is one of peculiar mathematical difficulty, but the general nature of a physically appropriate solution can be found. For example, with increasing distance beyond the neutral point, the velocity in the stream would be expected to be practically the general undisturbed velocity V, so that $y \to 1$ as $x \to \infty$. On the other hand, very near the sun the material will be falling in almost freely and thus have practically the parabolic speed, so that $y \to -\infty$ as $x \to 0$.

The region that is of most interest for the formation of comets is that extending inwards from the neutral point but otherwise at great distance from the sun, which means distances out to $\alpha G M V^{-2}$. In estimating numerical values for this, numbers have to be assigned to both α and V. The former, as has been mentioned, must lie between 1 and 2, and from the general instability of the accretion process, which emerges from the study of the full equations of motion with time-variation terms included, it seems possible that α might even vary during the motion in response to irregularities in velocity and density within the cloud. As for V, this will not only be expected to have a different value for each separate cloud that is encountered but may vary somewhat as the sun passes through it, again owing to slight internal motions within the cloud. Table 5.1 shows values of $\alpha G M V^{-2}$ in astronomical units for a series of values of α and V.

Table 5.1. Distance of neutral point $r_0 = \alpha G M V^{-2}$ (AU)

V (km/s^{-1})	$\alpha = 1\cdot 25$	$\alpha = 1\cdot 50$	$\alpha = 1\cdot 75$	$\alpha = 2\cdot 00$
3	124	148	173	198
4	69	83	97	111
5	44	54	62	71
6	31	37	43	49

Table 5.1 illustrates the large range of distance that can be involved in the process, and the correspondingly great range of initial periods of the motion. The actual time of infall in the accretion stream is in fact longer than that of free fall, because of the presence of the term AV in eqn. (5.18), but the periods associated with the distances shown range from about 60 years to 1000 years. As has been seen, however, once the comet is formed and set into almost parabolic orbital motion, the period of successive returns is immediately subject to large changes corresponding to $\delta(1/a) \sim (2\cdot 5 \times 10^{-4})$ AU^{-1}, as a result of the action of the planets; and even comets that are initially tightly bound to the sun, that is with small values of $\alpha G M V^{-2}$,

will eventually be brought to have periods of the order of 10^4 to 10^5 years or longer, or even be ejected altogether.

It has next to be shown how comets can form within this stream. Whereas in the original cloud the density is so low that self-gravitation is utterly negligible, the increase of density near the axis is so great that there is immediately the possibility of such action producing aggregations within the stream. There would be two opposing forces at work: first the self-attraction of a segment of material within the stream, and secondly the differential action of the sun arising from the increasing solar attraction with decreasing distance. In the idealized analysis of the accretion process, based on equations (5.17) and (5.18), the former force is ignored, so it is necessary that self-attraction should proceed sufficiently slowly not to alter sensibly the representation of the motion by a continuous accretion stream. This does in fact hold, as will be shown below. To examine the present point, let us consider a small element of the accretion stream of length $2d$ situated near the neutral point. Since its total mass will be $2md$, the self-attraction at either of its ends will be an acceleration of order $G(2md)/d^2$, while if the sun is at distance R its action opposing this is $\delta(GM/R^2)$, where $\delta R = d$. Thus the self-attraction will be stronger if

$$Gm/d \geqq GMd/R^3, \quad \text{or} \quad d \leqq (mR^3/M)^{\frac{1}{2}}. \tag{5.25}$$

This shows, as would be expected, that condensations are most likely to form where m is greatest and R is greatest, that is at the outermost part of the stream. In order to arrive at some numerical estimate of d, it is necessary to adopt precise values for α and the velocity V. For instance, if we take $\alpha = 1.5$, $V = 3$ km s^{-1}, then

$$d \leqq 1.3 \times 10^{10} \text{ cm}, \tag{5.26}$$

while for other values d varies as $\alpha^{\frac{5}{2}}\rho^{\frac{1}{2}}V^{-5}$.

The corresponding upper limit of mass for a comet so formed is clearly, by (5.25),

$$2md = 2(m^3R^3/M)^{\frac{1}{2}}, \tag{5.27}$$

and the most massive comets can thus be expected also to form at the outermost part of the stream. For the same numerical values of α and V, this gives, the general order of the mass of the comet as

$$2md \leqq 8 \times 10^{17} \text{ g,} \tag{5.28}$$

while for other values of the parameters this mass varies as $\alpha^{9/2}\rho^{3/2}V^{-9}$, the strong dependence on α and V suggesting that considerable range of masses may result from the process.

The line-density near the neutral point for these same numerical values is, from (5.21),

$$m = 2\pi\rho\alpha^2 G^2 M^2 V^{-4} = 3{\cdot}1 \times 10^7 \text{ g cm}^{-1}, \tag{5.29}$$

but the associated volume-density will be small, for it has been seen in (5.15) that the transverse width of the stream is of order 2.10^{10} cm, and this indicates an average volume-density of only about 10^{-13} g cm^{-3}.

These numerical values for the dimensions and mass of a comet are plainly of the right order, though the fact that α, ρ, and V may be subject to considerable choice shows that some range of dimensions, and a much greater range of masses, will result according to the circumstances of the encounter. Only about 2 per cent of encounters with dust-clouds would take place at a speed as low as 3 km s^{-1}, but an increase of density by a factor of 10 would be offset by a velocity of about 4·5 km s^{-1} and about five per cent of velocities would not exceed this value. On the other hand, little is known for certain of cometary masses, and many comets may well be much less than 8×10^{17} g in mass. In view of the uncertainties associated with the various parameters, and the high powers to which some of them enter, it is perhaps remarkable that results so similar to cometary values emerge from analysis of the accretion process.

Before continuing we may pause to examine at what rate the contraction of the segments into separate clouds proceeds. We can thus make certain that it does not happen so fast as to reduce the effectiveness of the accretion process. First, it is easily shown that a mass $2md$ of linear extent $2d$ would contract to half its size, if unhindered by the solar disruptive influence, in a time of about $0.64\,dG^{-1/2}m^{-1/2}$, and since $d \leq (mR^3/M)^{1/2}$, this time of contraction is less than $0.64(R^3/GM)^{1/2}$.

On the other hand, the time of fall to the sun from distance R would be $(\pi^2R^3/8GM)^{1/2}$, or about $1.1(R^3/GM)^{1/2}$. This time is also equal to about $1.1\alpha^{1/2}GMV^{-3}$. It follows that self-attraction within the stream would be great enough for it to divide into separate segments before reaching the sun, but the time of contraction is of the same order as the time required to build up the steady state, namely a few times GM/V^3, which is the characteristic time associated with the accretion process. Thus, contraction into these segments will proceed sufficiently slowly for the stream to approximate to the steady-state distribution at all times, especially in the outermost part of the stream, where the density will be greatest. The rate of contraction at this stage will probably be extremely slow, for not only would it be hampered by the solar effect, but in a time of this order a contracting segment that has achieved individuality by separation away from the stream at its ends would receive a supply of material from the sides comparable with its own mass, and this would presumably tend to disperse the segment and so further hinder contraction.

It is an essential point of the accretion theory that the comet can hold together by self-gravitation when it first forms, and it is clearly the presence of the factor R^{-3} in the differential disruptive effect of the sun that renders this possible despite the small masses and large sizes of comets. A further important point arises in this connexion : even with $V = 3$ km s^{-1}, the time of contraction of a comet would be less than 200 years, and once a comet is formed its mass and size will be such as to preserve this intrinsic period. But planetary action may soon change the orbital period to a very large value in excess of 10^4 years, and send it into a path for which R is now much greater than the original accretion value of order GM/V^2. Accordingly the self-gravitation of the comet would then at times far outweigh the solar disruptive influence. Moreover, because of the large size of the comet, the motion of contraction would take the form of an oscillation in which a particle starting at a given radial distance moved through the central part out to a corresponding distance diametrically opposite the original position. This would happen for all particles except in so far as their motions might be damped by collisions with other particles and by viscous damping by gas, itself produced by collisions. Despite the low density of the comet, it cannot be thought that the probability of these collisions is small; and it seems clear that a long-period comet can pull itself together into a much more compact overall volume, at least while pursuing the aphelion side of its orbit. On the other hand, immediately after formation, a comet has first to approach the sun, and, as will be shown below, at this stage extension of the comet along its orbit is likely to occur. But the time spent on the perihelion side of its orbit is at most a few years, whereas for a period of 50 000 years, say, a comet could undergo some 250 internal oscillations, and would probably completely overcome any disruptive effects produced while near the sun so long as these were not great enough to disperse the comet altogether. We shall return to these considerations below when we consider tail-production by comets.

It is even possible to arrive at an estimate of the total number of comets formed during passage of the sun through a cloud of given dimensions. For example, if the path of the sun through the cloud measured D cm, the time of passage would be D/V seconds, and the total amount of material captured in such a time would be, by (5.5) and (5.21),

$$\alpha G M V^{-2} . 2\pi\rho G M V^{-1} . D V^{-1} = 2\pi\alpha\rho G^2 M^2 D V^{-4}. \quad (5.30)$$

Using the expression on the right-hand side of equation (5.27) for the mass of a comet, the total number that could be formed from this mass is accordingly

$$D V^5 / (2^{1/2}\pi^{1/2}\rho^{1/2}\alpha^{1/2}G^{5/2}M^2). \quad (5.31)$$

As a numerical example, suppose $D = 0.3$ parsec $= 10^{18}$ cm, together with the values $V = 2 \text{ km s}^{-1}$, $\rho = 10^{-24} \text{ g cm}^{-3}$,

$\alpha = 1.5$, as adopted before; the total amount of dust gathered in by the sun is then about 2×10^{25} g, and the value of (5.31) is therefore as high as 5×10^7 comets. Curiously enough, this number increases with V and decreases if ρ is increased, since it is proportional to $V^5 \rho^{-\frac{1}{2}}$. Evidently the reason is that the mass of an individual comet depends on $V^{-9} \rho^{\frac{3}{2}}$, whereas the total mass collected in varies as $V^{-4} \rho$.

It cannot be supposed that this large number would all remain as permanent members of the solar system. To begin with, the periods will be short, of the order of a few hundred years, and the rate of expulsion by planetary action will proceed very rapidly. For $\alpha = 3/2$, $V = 3$ km s^{-1}, for instance, the initial value of a would be 74 AU; the binding energy (in units such that it is $1/a$), would then be 1.35×10^{-2}, while at each approach an energy change of order 0.05×10^{-2} could occur. Thus, in the early stages an extremely rapid sorting-out of comets will happen, as the periods disperse: those becoming more tightly bound will be swept into the sun, and the others will achieve increasingly long periods until the energy changes become such that they will bring about ejection of the comet concerned from the solar system.

It is also possible to form some estimate of the number of such passages through dust-clouds that the sun is likely to have undergone. The volume of the galaxy at present occupied by dust-clouds is estimated to be of the order of about one-tenth of the whole. If an average situation is regarded as representable by a path of length 10^{18} cm and a speed of 3 km s^{-1}, the time of passage is about 10^5 years, and accordingly on average the formation of a group of comets by a single such passage would occur every million years or so. Now only a few per cent of encounters are likely to occur at speeds less than 5 km s^{-1}; so allowing for this, perhaps once every 2×10^7 years there is a really successful comet-producing encounter. But it has been seen that there are a number of comet-groups, of the order of fifty perhaps, or more, headed by the famous sun-grazing group with eight or nine known members to date, but mostly containing five, four, three, or even only two members. The origin of fifty such groups would then go back 10^9 years in time, and it seems doubtful whether many comets older than this could survive, for they have to contend not only with the possibility of dynamical ejection, but, if they remain in the system, also with the physical dissipation and mass-loss that must occur at every perihelion return. As pointed out above, with an average period of 50 000 years, a loss of mass of only one part in a thousand would mean a reduction of total mass by a factor $1/(6 \times 10^8)$ if maintained for 10^9 years, and there would not be much left of the comet long before then!

An especially important feature of the accretion process from an observational point of view is that it endows the newly-formed comets with orbits that are almost parabolic, but nevertheless elliptic. The absorption of the dust into the accretion stream destroys the energy corresponding to the transverse component of velocity, and it is this that reduces the hyperbolic interstellar energy of the dust and enables the sun to effect capture. And as has been seen, it is because the eccentricity e is so near to unity that planetary action can bring about such large changes in a. But it has yet to be considered why the accretion stream does not simply flow into the sun. If the theory is to account for comets, then some further factor must be involved capable of giving the newly formed comets angular momentum about the sun. At first sight it seems possible that the attractions of the planets, Jupiter in particular, might achieve this. Calculations show, however, that a particle released at great distance at rest relative to the centre of mass of the solar system, which is the point towards which it would initially begin to fall as the mean centre of attraction, would have an eventual path that would nevertheless take it very close to the centre of the sun, well within the actual bodily radius of the sun. It seems unlikely that even the small angular momentum of the sun-grazers could be explained in this way, while at the other extreme many long-period comets are known with values of $q = a(1 - e)$, which effectively is a measure of the angular momentum, as high as 1 AU; and there are a few even in excess of 3 AU. Whether the absence of higher values is real or due entirely to selection effects is a question at present unsettled, but this apart the primitive comets must be endowed with some angular momentum. A mechanism is not far to seek, for the dust-clouds are actually seen to be highly irregular in shape and presumably also in density. The amount of material arriving at the accretion axis from different directions may not therefore be the same, and the net angular momentum brought to the accretion-stream may not cancel out precisely to zero. It is to be noticed that the accretion-axis itself is in a direction defined by the velocity and not by the amounts of material arriving at it from different directions. At great distances from the sun near the neutral point the effect may only be very slight, but as the material flows inwards, if it possesses angular momentum about the sun, it will move off the axis and will be no longer or very little affected by material coming in laterally. Thus it seems possible that comets first achieve their identity at the outermost part of the stream, and then, as a result of slight angular momentum, shear off sideways to go into elliptic motion round the sun.

The foregoing analysis of the accretion process has been based on the steady-state solution, which is difficult enough in itself, but to what extent this adequately represents what might happen for an actual time-varying system is an open question. If the cloud is irregular in distribution of density, and if also the relative velocity varies slowly, then clearly time-dependent terms must be involved, and it can only be conjectured that in some circumstances accretion might be more effective than is suggested by the steady-state solution and in others less effective. The complete equations even for line-accretion corresponding to (5.17) and (5.18) present exceptionally great mathematical difficulties that at present stand between us and the full understanding of the accretion process.

12—M.S.S.

TAIL-PRODUCTION BY COMETS

A theory of the origin of comets would scarcely be adequate if it failed to give a picture of their structure that could account for the peculiar activity of gas-release and tail-production. That this occurs and increases in intensity as the comet approaches the sun has suggested to many that it must be connected with solar heat. While solar heat may have some influence on comets, it is found that their activity is not directly related to decreasing solar distance. For example, comet Humason (1962 VIII) has a perihelion distance in excess of 2 AU (and aphelion at over 400 AU) and so never comes within the distance even of Mars from the sun, yet it shows intrinsic tail-production far stronger than many comets coming very much closer. The great daylight comet Ikeya–Seki (1965 f), which proved to be a member of the sun-grazing group, was well inside the orbit of Mercury and within a few solar radii of the sun before even the minutest amount of tail-production was observable, but after passing perihelion the tail was sufficiently intense to be easily visible even in the dawn sky. Solar heating of comet Ikeya–Seki long before perihelion would have been hundreds of times stronger than it ever is on comet Humason, while the tail-production by Ikeya–Seki at this stage still remained so negligibly small as to be detectable only by long photographic exposure (Plate 5b).

It has been seen that a comet first achieves its identity as a result of its self-attraction overcoming the disruptive tendency of the solar attraction, perhaps not by any great margin. Owing to the generally unstable nature of the accretion process, comets are unlikely to have regular shapes, but for the purposes of calculation, the comet may be assumed to have spherical form, and it will be supposed for numerical purposes that the mass m_c and radius r_c are given representatively by

$$m_c = 10^{18} \text{ g}, \quad r_c = 1\cdot5 \times 10^{10} \text{ cm.} \tag{5.32}$$

Since the distance R from the sun at which the two effects will be roughly equal is then given at once by

$$\frac{m_c}{r_c^2} = \frac{2M}{R^3} r_c,$$ (5.33)

this leads to

$$R = 2 \cdot 4 \times 10^{15} \text{ cm} \simeq 150 \text{ AU}.$$ (5.34)

The greatest distance from the sun at which any comet has been observed is just over 11 AU, for 1927 IV, and the vast majority are not detected until $R < 5$ AU. Because of the presence of the factor R^{-3} on the right-hand side of (5.33), the gravitational action of the sun when the comet is at such distances will thus far exceed the self-attraction of the comet. Thus at 4 AU the ratio of the right-hand side of (5.33) to the left-hand side is as high as 6×10^4; at 1 AU the solar action is greater by more than 4×10^6; and if the comet were a sun-grazer, then at perihelion (where $R = q = 10^{11}$ cm) the ratio would exceed 2×10^{13}. Since the force given by the left-hand side of (5.33) refers to particles at the outer part of the comet, the self-attraction on the interior particles would be even less, and the present ratio greater still.

The dynamical implication of these figures is plainly that while the comet is within planetary distances of the sun, its self-gravitational influence is negligible, and hence that *each particle going to form the comet will describe a separate independent orbit about the sun.* If the distance $R = 150$ AU arrived at above corresponds roughly to the end of the minor axis of the orbit (where $r = a$), the period would be about 1800 years, and the comet would spend almost nine-tenths of this time pursuing the outer half (beyond the minor axis) of its elliptic path; during this part of the period the self-attraction would predominate and the cometary particles would all be brought to have practically the same period about the sun. The orbital speed of the comet, supposing $q = a(1 - e) = 1$ AU for numerical purposes, would be about $1 \cdot 4 \times 10^4$ cm s^{-1} at aphelion, and will have reached about $2 \cdot 4 \times 10^5$ cm s^{-1} before the differential attraction of the sun begins to take over. On the other hand, the internal speeds produced solely by the self-attraction of the comet would be only of order 1 cm s^{-1}, and with a uniform steady spherical distribution the motion would be simple harmonic (with period about 10^3 years). Only near the centre of the distribution would such a value of the velocity be attained, and also for only a few particles would the speed be in much the same direction as the orbital velocity vector.

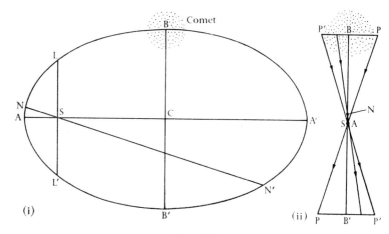

FIG. 5.3.

(i) Diagram showing in plan the standard orbital plane of a comet. A particle lying out of this plane when the comet is at B (the end of the minor axis, for example) will move in a plane passing through S and meeting the standard plane in a line NSN'. At N and N' it will cross through the standard central plane. For almost any distribution of the nodes N', the points N will cluster strongly near A.

(ii) Shows the planes of motion in elevation as viewed from a great distance along the major axis. A particle starting at P' (near B) will cross through the standard plane BB' at or near A. This will occur for every particle of the comet at some point between L and L', and for most this will be close to A.

Accordingly, only a minutely small proportion of the particles of a comet would have (osculating) period differing from the general orbital period by even as much as 1 part in 10^5 at the stage when the solar attraction takes control of the motion, and this would be long before the comet would have become observable from the Earth. Moreover, even if any such particles moved away from the main body of the comet, there would be no reason to expect them to be observable.

For the sake of definiteness, it may therefore be supposed that on the inner half of its orbit, from one end of the minor axis to the other, the self-attraction of the comet is entirely negligible. We then have to consider how such a swarm of particles, all moving with the same orbital period, would behave as it approached the sun. For this purpose, let us single out some imaginary particle at the centre of the distribution and utilize its path to define a standard orbit. Then all other particles will be pursuing adjacent orbits only slightly different from this standard one, except that all are to be regarded as having the same period and hence the same value of a. Most important of all, each particle will describe an elliptic path lying in a plane through the centre of the sun S as focus (Fig. 5.3). The simplest orbits of these to envisage will be those of particles initially on the line at B perpendicular to the standard plane, for their paths will be obtainable simply by a slight rotation of the standard orbit round the major axis AA'. For these, every particle P above the standard plane will follow a path intersecting this plane at A, and two particles P and P' initially on opposite sides of this plane will have a relative velocity when at A equal to

$$\frac{PP'}{b} \text{ (orbital speed at perihelion)} = \frac{PP'}{q} \bigg/ \sqrt{\left(\frac{GM}{a}\right)}. \quad (5.35)$$

With the numerical values already adopted for a typical comet, namely $PP' = 3 \times 10^{10}$ cm, $a = 150$ AU, this relative speed is

$$5 \times 10^2 \, (q_{AU})^{-1} \text{ cm s}^{-1}, \quad (5.36)$$

where q_{AU} is the perihelion distance in astronomical units. For comets of different dimensions this speed varies simply as r_c.

Similarly an adjacent orbit may be obtained by a small slight rotation about the chord LSL' perpendicular to the major axis. Viewed in the direction along the major axis, the extreme path would appear as in Fig. 5.4 (p. 172). The diagram much exaggerates the size of LL' compared with BB', for in fact LL'/BB' is equal to $(2q/a)^{1/2}$, and so may be of the order of 10^{-2} or less.

Figs. 5.3(ii) and 5.4, which relate to particles farthest from the standard plane when the comet is at B, both show that the comet will *contract* in directions transverse to the orbit as it

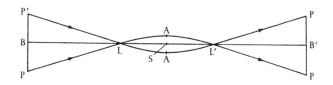

Fig. 5.4. Paths of particles in slightly inclined orbits through LSL' as seen from great distance in the direction of the major axis. (See also Fig. 5.3.)

approaches perihelion. The degree of this contraction will be settled by the second case, and so will be by a factor

$$\frac{AS}{SC} = \frac{q}{ae} \simeq 1 - e, \tag{5.37}$$

where lengths perpendicular to the orbit are concerned. This is in agreement with the general result of observation that comets appear to contract as they approach the sun.

A third way of obtaining an adjacent orbit is obviously by a slight rotation of the standard orbit in its own plane by a small angle lying between $-r_c/a$ and $+r_c/a$. The consequent relative velocity at perihelion for the two extreme paths so obtained is readily found to be

$$\frac{2r_c}{a}\left(2GM/q\right)^{1/2}. \tag{5.38}$$

Plainly this is smaller than (5.35) in the ratio $\sqrt{(2 - 2e)}$, and it follows that this contribution to the relative speed of particles within the comet will not usually be important since $e \sim 1$.

Accordingly, the orbit of an individual particle can be regarded as obtained by a combination of the first two rotations, which will be equivalent to a small rotation of a standard orbit about some line NSN' (Fig. 5.3(i), p. 170) in its plane. For such an orbit, the particle will pass through the standard plane twice per revolution, namely at N and N', which means that the comet must 'turn itself inside out', as it were, twice per revolution: once on the perihelion arc LSL', and once on the aphelion arc $L'AL$. Thus, other effects apart, all those particles which when the comet is at B are above the standard plane will cross down through it, at points such as N, and those below the plane will cross up through it during the time the comet travels the arc LAL'. It is to be noticed that for any ordinary distribution of the nodes N' on the aphelion side of the orbit, where the comet spends almost all its existence, the distribution of the nodes N on the perihelion side is very strongly concentrated near A, and the greater the eccentricity of the orbit the more this will tend to occur. Thus, the great majority of the particles will cross through the standard plane close to the point A. For comet Encke, with $e = 0.846$, almost 90 per cent will cross between $\pm 60°$ of A; while for Halley, with $e = 0.967$, this percentage will actually cross within $\pm 20°$ of A; and the nearer e approaches 1 the smaller will be the corresponding range of angle.

It is to be noticed also that the time spent on the perihelion side of the orbit is always only a minute part of the whole period. If P is the total period, this time of describing the arc LAL' is given approximately by

$$0.6(1 - e)^{3/2} P. \qquad (5.39)$$

Thus, for Encke, with $e = 0.846$, $P = 3.3$ years, the time is about 1.4 months; for Halley, with $e = 0.967$, $P = 76$ years, it is about 3 months. But for comets with $1 - e \sim 0$ and of small perihelion distance, the time on the perihelion side may be only a matter of hours, while the aphelion side takes thousands of years. It is also of importance to the development of a comet that the speed on the perihelion side is far greater than on the aphelion side. At these precise points, A and A', the ratio of speeds is in fact $(1 + e)/(1 - e)$. At perihelion, comet Encke moves at about 70 km s^{-1}, and Halley at 60 km s^{-1}, but a sun-grazing comet may reach a speed in excess of 500 km s^{-1}.

When the comet crosses through itself in accordance with these ideas, it is plain that there will be the possibility of collisions between some of the particles. It is in this way that pulverization into smaller particles will occur together with the release of occluded gases, and possibly the formation of gases from compounds within the particles. But before considering this in any detail, another effect to be noticed is that as the cometary swarm speeds up in approaching perihelion it will extend in the general direction of the orbit. For if two particles are describing exactly the same orbit but are slightly separated in time, their distance apart will vary in direct proportion to the instantaneous orbital speed. As between perihelion and the end of the minor axis, the ratio of the speeds is $(1 + e)^{\frac{1}{2}}$: $(1 - e)^{\frac{1}{2}}$, and for a long-period comet this factor will usually be considerably greater than 10. On the other hand, for short-period comets such as Encke and Halley, at no time will their self-attraction exceed the solar influence, and the ratio of greatest to least speeds will be $(1 + e) : (1 - e)$, corresponding to perihelion and aphelion. For Encke this ratio is about 12, and for Halley about 60.

But it does not follow that the comet will necessarily be *observed* to increase in length by such a factor. For where a swarm of particles is concerned, increase in separation could result in an apparent decrease in size, since the outer parts would tend to become even more diffuse. Secondly, the dimension of a comet here involved is that along the orbit at and near perihelion, and the Earth and comet would need to be specially situated for this to be observable and also for comparative observations to be possible with measures made at some other part of the orbit. If only minute solid particles were involved, the extension along the orbit might thus remain always undetectable. However, it is of interest here that the great daylight comet 1882 II passed so close to the sun that its constituent particles must almost certainly have been completely gasified, and that shortly after perihelion it was in fact observed as a brilliantly luminous streak extending in the direction of the orbit.

Clearly in this crossing-over from one side of the standard plane to the other there is, as already mentioned, the possibility of some of the particles undergoing collisions. The effect of these collisions will be twofold: first, the small particles of a comet are likely to have very weak structures if they have formed from gas suddenly released in interstellar space, and collisions between them would fracture them into a large number of smaller particles. Thus, one particle of dimensions 10^{-3} cm could produce 10^6 of dimensions 10^{-5} cm, and these would have 10^2 times the cross-sectional area of the original single particle. In practice, a collision would produce a distribution of particle sizes, and if a single particle of dimensions 10^{-3} cm produced a range from 10^{-4} cm down to 10^{-7} cm, an increase of surface area by a factor of order 10^3 could result; and even if only a small proportion of the total mass were involved, an increase by 10^2 in surface area for the whole comet could clearly result. There can be little doubt that this is the explanation of the great increase in power of ordinary reflectivity that comets exhibit on their approach to the sun.

Undisturbed dynamical motions would be symmetrical about perihelion, but as has been seen collisions become more probable once they begin, and accordingly the brightening of a comet through the present effect would be expected to be greatest at some time after perihelion. This in fact is found to be the case for many comets: in particular those with perihelion distance large enough for complete vaporization of the particles not to occur. Even for a comet with large q, such as comet Humason with $q = 2 \cdot 133$ AU, the relative velocity of particles on the perihelion side of the orbit would still be of order 10^2 cm s^{-1} and sufficiently great to shatter weak structures. But no serious effects of solar heating could be expected at such a distance from the sun, though some intense local heating on the particles through collisions could occur.

The second effect likely to result from collisions is the release in gaseous form of volatile substances. When a meteorite strikes the lunar surface and is brought to rest, its kinetic energy is not immediately distributed through the whole mass of the moon, but is absorbed only locally, as it were, by high-temperature effects. Similarly, even when two small particles collide, their relative kinetic energy must be communicated initially to those molecules through which contact is first made, and temperatures corresponding to thermal speeds equal to the relative speeds will be generated locally at once; but as penetration increases, adiabatic compression of the initially heated volume will take place and raise the temperature still further. A relative speed of the order of 10^4 cm s^{-1} would mean a kinetic temperature of a few hundred degrees for heavy molecules; and so local temperatures of the order of 10^3 °K or higher could result from compression. It is a matter of common experience that flints rubbed together at very low speeds, of only a few centimetres per second, will produce flashes of light (implying ionization) and gaseous emission from the stones. The same must hold for low-speed collisions, and it is to be expected that at the lowest speeds the most volatile constituents will be driven off, while with increasing speed less volatile substances will become expelled, and so on. It is not to be expected that all parts of the comet will take part in collisions at every return, and not only will freshly exposed surfaces result from collisions but the collisions themselves will penetrate inwards and reach a depth where occluded gases can hitherto have been retained.

It is even possible that this release of gas within the whole general volume of the comet could result in the formation of something akin to a nucleus, for automatically such gas must expand outwards, and to do so would form a kind of centre from which the expansion took place. On occasion more than one such centre might form temporarily, giving the appearance of more than one nucleus, and if gas-production were very limited no nucleus might appear at all. If this suggestion is correct it could explain why the nucleus, when present at all, is a transient phenomenon, and as may be recalled its extent is also very difficult of measurement.

In the foregoing calculations, the comet has more or less unavoidably been regarded as a spherical distribution of particles, but in reality the limiting boundary to which particles extend may be highly irregular. Moreover, this limit is likely to extend considerably beyond the observed limits of the coma. Just how far is uncertain, but it could well be quite a large factor, judging by the extents of meteor streams, higher perhaps in the direction along the orbit than in the transverse direction. The relative speeds of colliding particles could therefore be proportionately greater than those that would be indicated by the mere size of the coma, and it is possible that collision speeds in excess of 10^5 cm s^{-1} can occur for large comets of small perihelion distances.

Where the group of sun-grazing comets is concerned, another highly interesting development must also take place, for the black-body temperature due to solar radiation at a distance of a mere half radius of the sun from its surface would be in excess of 4000°K, and complete vaporization of small particles would occur. Thus, such comets must be regarded as becoming entirely gaseous for the few hours they spend near perihelion. But this does not mean that the comets would dissipate by evaporation: the surface of the sun is extremely cool judged by the thermal velocities there in comparison with the speeds associated with its gravitational field. At and near perihelion the orbital velocity of the comet would be about 500 km s^{-1}, but the thermal speed due to solar heating would be only 1 or 2 km s^{-1}, and negligible expansion of the comet could therefore occur, and would in any event be difficult to disentangle from other changes of shape occurring for the dynamical reasons already described. As soon as such a comet had receded to more than a few solar radii, the temperature within it would have fallen to such a value that the material would condense again into small solid particles; and then at far larger distances self-gravitation would enable the comet to pull itself together again into a fairly compact form.

It is clear from these considerations how for different comets tail-production can occur at such different distances from the sun. As already noted, the great comet Ikeya–Seki (1965 f) showed almost no sign of any tail while still at only a few solar radii from the sun on its inward path. This would be simply because the perihelion side of the orbit was confined within a smaller distance still of only 3·5 solar radii from the centre of the sun. On the other hand, comet Humason exhibits strong tail-formation while always remaining at more than 500 solar radii from the sun. This results from the great difference in their perihelion distances, and plainly rules out solar heating as the principal cause of tail-production.

The production of tails by comets is a consequence of these developments. In the first place, it has long been considered established that particles of the order of the wavelength of light in dimensions will be repelled by solar radiation-pressure, and, with comminution of cometary particles occurring mainly on the perihelion side of the orbit, light-pressure will automatically select all those of appropriate size and expel them from the comet in hyperbolic orbits away from the sun. Where gas is concerned, selective absorption and re-emission by certain molecules, such as CO^+ for example, can produce large repulsive forces; and here again it will be just those substances for which the effect is large that will automatically be driven out of the comet and continue to be repelled from the sun. This does not exhaust the possible causes of the acceleration of the tail material. In recent times, the discovery of the solar wind has led to attempts to account for cometary tails by its action. But the momentum associated with the solar wind is at most only about 10^{-5} that of solar radiation (and probably a good deal less); some highly effective linkage, through magnetic fields increasing the effective cross-sectional area of the molecules, would therefore be needed if a related process is to prove adequate. There is also the suggestion, first made a century ago but not yet adequately investigated, that excess evaporation on the sunward side of dust-particles produces a net recoil away from the sun, rather after the manner of rocket-propulsion. This, of course, could apply only to dust-particles, not to gas, but it remains possible that radiation pressure supplies only part of the strong repulsive forces that undoubtedly exist.

To simplify the discussion it has been supposed in much of the foregoing that all the particles of the comet maintain precisely the same orbital periods. But collisions between particles will dissipate energy, and the resulting particles if moving solely under the sun's attraction will therefore not have quite the same periods as before. They would therefore move slowly away from the main mass of the comet: those that lose energy will have their periods reduced, and so will move on slowly ahead of the comet, practically in the orbital direction; those gaining energy will slowly lag behind. For long-period comets the effect might well be overcome on the aphelion side of the orbit; but for short-period comets self-attraction can never rise to importance, and particles so ejected will be lost forever from the comet. For ejection speeds as low as a few centimetres per second, distribution right round the orbit for comet Encke, for example, would require less than 10^5 years; for comet Halley the time would be less than 10^6 years. There is also the drag effect of solar radiation, which acts with different strength on particles of different sizes, and more strongly the smaller the particles. This also will operate to send the smallest particles on ahead of the main mass of the comet.

Such ejected particles can evidently be identified with the meteor-streams known to be associated with several short-period comets whose orbits are intersected, more or less, by that of the Earth. It is a further indication that comets consist entirely of small particles that no object has ever penetrated to ground level during any meteor display, however intense. The dimensions of tektites and small meteorites indicate that, in order to reach ground level, an object entering the atmosphere as cosmic speed must initially be at least a few centimetres in size. During a great meteor-shower, many thousands of millions of meteors enter the Earth's atmosphere, but there has been no recorded instance of a single meteoritic fall during the time of any such shower.

Where other theories of comets are concerned none go so far as to show how they have originated as comets. The icy-conglomerate model, already referred to above, postulates a detailed physical structure for the nucleus with the object of accounting for the observed cometary phenomena, but it does not provide any explanation of how the nuclei have come to develop within the solar system. As has been seen, the suggestion that comets originated at the same time as the planets leads to insurmountable difficulties associated not only with their expulsion from the solar system but with the loss of mass at each return to the sun. Some possibility of avoiding this mass-loss has been explored by the revival of the ancient notion that a vast assembly of comets exist in a spherical shell at great distances from the sun. According to this theory, the shell contains 10^{11} comets in all, most of which remain always at distances of 10^4 to 10^5 AU from the sun. It is suggested as part of the theory that those that *are* seen at planetary distances represent a comparative few that have been deflected in by the perturbations of passing stars.

In recent times definite evidence of this shell has been claimed to have been found by considering the distribution of numbers of comets observed for different values of the binding-energy. Thus if the number of comets $n(E)\,dE$ in a small range of energy is plotted against the energy $E = 1/a$, it is found that there is a strong peak at small values of $1/a$, and *hence*, it is claimed, there must be a high volume-density of comets subsisting at large values of a itself. But the argument is transparently fallacious. If it is desired to demonstrate a space-density of any population, then number *per unit volume of space* must be plotted against distance, or at least against something *increasing* monotonically with volume itself. In the present case, the peak for small values of $1/a$ means only that there are numerous comets with large values of a; but as a can take all values up to infinity there is ample room for a large number without implying any high density at great distance. It is further maintained that the plot of numbers of comets against $1/a$ is the natural way to proceed because $1/a$ is proportional to the binding-energy; but on similar grounds it could equally well be argued that they should be plotted against $a^{3/2}$ since this

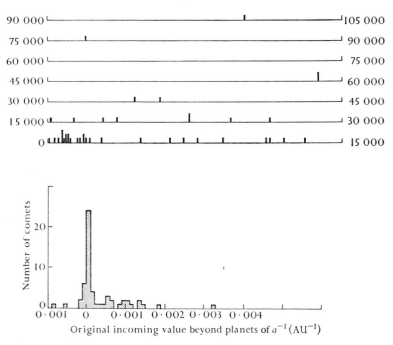

FIG. 5.5. Original values of semi-major axis (in AU) for 57 long-period comets. Values taken from Galibina, *Bull. Inst. theor. Astron.* **9**, 496 (1963–4).
[Not plotted: three with $a^{-1} = 7 \times 10^{-6}\,\text{AU}^{-1}$, three with $a^{-1} = 2 \times 10^{-6}\,\text{AU}^{-1}$; six with negative $a = -1170, -1700, -6900, -9100, -15\,000, -42\,000\,\text{AU}$.]

is proportional to the period, and for this in fact the peak would be found to be non-existent. However, as aforesaid, to demonstrate an excess density in space, the plot must be made against volume elements at progressively increasing distances, that is against $a^2 da$; but even if the factor a^2 is omitted and all comets with the same a regarded as contributing to a simple line-density in a to $a + da$ there is no sign whatever of the alleged peak, and all one gets is a number of points spaced out irregularly with no sign of any congestion (Fig. 5.5). Making allowance for the factor a^2 would spread them out even more sparsely. Thus the very line of argument that has been regarded as demonstrating the existence of the shell of comets when properly carried out demonstrates its non-existence.

6 Tektites

TEKTITES are small glassy objects that are found widely scattered at a few regions of the Earth's surface. They are black in appearance and about the size of small nuts, though they vary considerably and take a great many shapes. The name is derived from the Greek *tektos* (molten) and was first put into use for the whole class by Suess in the year 1900, it having already been concluded that tektites must have passed through some earlier molten state. The general recognition that they represent some exceptional type of object dates back at least to 1787, and the possibility that they may have a cosmic origin seems first to have been proposed in 1897.

There are eight main areas in the world generally accepted to be genuine tektite-fields. The tektites present in each area to some extent have distinctive properties, but whether the Asian fields are all quite separate and independent in origin is not quite certain. A few other areas, in Colombia, Peru, and Libya, for instance, in which small vitreous objects have also been discovered, have from time to time been claimed to be tektite-fields, but it remains highly doubtful if these can be so classified owing to the considerably different properties, and there are still other areas now definitely rejected from the list. Table 6.1 (p.184) gives the various accepted areas together with the special names assigned to the tektites found there, and also an estimated figure for the total number collected in each field to date.

Naturally enough, the earliest records are European: they

13—M.S.S.

relate to tektites found in Czechoslovakia, and named after the Moldau river in Bohemia. In Australia and adjacent islands discoveries were first made as early as 1834, and examples were also found on the East Indian island of Billiton in 1836, but the remaining fields seem all to have been recognized in the twentieth century, though almost certainly specimens were come upon far earlier in every case. The dimensions of the several areas are all small compared with the Earth as a whole, though they vary a great deal. The largest, more than 5 million km^2 in extent, is that of the australites, which are found in a great many areas south of a line running across the continent roughly parallel to the 25°S-latitude line; practically no tektites are

Table 6.1. Known localities of tektites.

Area	Special name	Total number collected
Philippine Islands	Rizalites	500 000
Czechoslovakia	Moldavites	55 000
Australia and Tasmania	Australites	40 000
Indo-China and Hainan	Indochinites	40 000
Malaya	{ Malaysianites } { Billitonites }	7500
Java	Javaites	7000
Texas (and Georgia), USA	Bediasites	2000
Ivory Coast, West Africa	Ivory Coast tektites	200

found north of this line, though to the south the field extends to Tasmania and other adjacent islands. The indochinites are spread in a broad band nearly 1600 km long in a roughly north–south direction from Tongking to Cambodia. Most of the islands of the Philippines contain tektites, and some are richer than any fields elsewhere. The moldavites have been found in two main but connected areas, in southern Bohemia and Moravia, covering some 1200 km^2. The Texas field has been found to cover at least 120 km^2, though there is growing evidence that it may be associated with a field in south Georgia across the Gulf of Mexico.

The total number of tektites so far gathered from all these

fields must amount to something like two-thirds of a million. Although they have no value as precious stones, the growing interest in matters scientific is producing collectors anxious to procure them, and the rate of discovery may well increase. Their wide distribution over each area concerned, as for instance the whole of southern Australia and beyond, has been the source of much speculation, especially with the object of accounting for their distribution by known terrestrial effects such as stream-action and transport by ice; these seem certain to have been operative, if only to limited degrees. It is known that tektites have been used by some of the larger birds as gizzard-stones, and this clearly would lead to some transfer of them. It might even account for the *one* tektite found on the island of Martha's Vineyard in the north-eastern United States, though it could obviously also have been carried there by man. Even the aborigines, who used tektites as a primitive coinage, have been credited with having a hand in the matter, but it seems absurd to suppose that their actions alone are responsible for the dispersion actually found. The consensus of opinion among modern field-workers appears to be that most if not practically all tektites are found not far removed from where they originally fell.

Unlike the distribution of meteorite finds, the distribution of the tektite-fields themselves on the Earth's surface has no correlation with the distribution of population. No tektite-field is known in the whole Soviet Union (which occupies one-sixth of the entire land-surface) and none in South America or Canada. But within the Australian field itself, the density of finds does show some relation to the population-distribution, as would be expected if the whole vast southern area were more or less completely covered. The initial discoveries of tektites have been largely by chance, though natives and prospectors have collected them as curiosities. Prospectors have, it seems, occasionally found them in mines, but their special nature has been realized only by others. Tektites are in places actually to be found lying exposed on the free surface where the terrain is such that vegetation is absent, and wind and erosion remove the surface material, as in desert areas. But they are usually found loosely buried within a metre or so of the surface in gravels, clays, and sands, and are often turned up in digging operations and ploughing. Less frequently they are found at depths of 5 to 10 m in alluvial deposits. They are also commonly found at various depths resting on bedrock below such deposits, though they themselves bear no discoverable chemical relationship to the materials in which they are embedded. The extent to which tektites may occur at far greater depths elsewhere in the world is not known, presumably because little exploration in depth has been carried out. In this connexion, it may be noted that the combined area of all known tektite-fields amounts to only about 5 per cent of the land-surface of the Earth.

The tektites are quite haphazardly distributed over any sizeable area within a field. In part of Cambodia, for instance, as many as twenty occur within 1 m² at some places, while close by none are to be found in more than 20 m². At one find in Santa Mesa (Philippines) over 200 were located in an area as small as 5 m², while in Australia the greatest concentration found on the surface is about one in 200 m². It has been estimated that the total number of australites in the entire field may approach 10 million, and with an average weight of a few grammes each this would represent 10 000 kg or more by weight for them all.

When viewed by ordinary reflected light, most tektites appear to be jet black, but they are mildly translucent in varying degrees; some, particularly the moldavites, then show a greenish hue, and others a brownish tinge. Their surfaces are often strongly etched through weathering effects, in general to a lesser extent the more deeply buried they are found, doubtless owing to the protection afforded by the covering. Tektites are brittle objects, and when freshly broken the newly exposed surfaces have a brilliant dark lustre.

A remarkable property that almost certainly must be of special significance is that huge numbers of minute particles are found to be embedded within the general matrix that forms the bulk of the tektite. Most of these particles are in the size-range 0·14 to 0·03 mm but they range down to microscopic sizes. Several hundred per cubic centimetre of the larger particles are present, but in total they occupy less than 1 per cent of the entire volume. The particles consist mainly of non-hydrous silica and quartz, and are usually aligned with the solidified flow-structure within the tektite, but minute metallic spherules are also to be found.

Tektites have compositions somewhat similar to glass, but the refractive index of their material is usually very near to 1·5, compared with the value of about 1·68 for artificial glass. The specific gravity ranges from about 2·3 to 2·5 g cm^{-3}, mainly according to the proportion of silica, which varies from about 68 per cent for the densest to 81 per cent for the least dense specimens. The weights of the standard examples of tektites usually exhibited range from 1 to 100 g, but the extremes are from 0·06 g for the tiniest recognized to 3200 g (about 7 lb) for the largest yet found, which in fact is an indochinite. The largest known specimens recovered in the remaining fields are as follows: rizalites 1070 g, javaites 750 g, moldavites 500 g (Bohemia) and 235 g (Moravia), malaysianites 464 g, australites 218 g, bediasites 91 g, and Ivory Coast tektites 79 g. Tektites have no magnetic properties.

Table 6.2 (p. 188), based on values given for samples of all tektite-fields by G. Baker, shows the contained percentage by mass of the principal chemical constituents of an average tektite, and also the maximum and minimum quantities found. In addition there are trace-element contents of a few hundred parts per million of numerous other elements, the principal ones being barium, chromium, lithium, nickel, rubidium, strontium, and zirconium. The presence at less than 10 parts per million of lead and uranium is nevertheless important because of the possibility of age-determinations.

The shapes of tektites are extremely varied, though they are repeated very frequently in different examples, and a large number of everyday terms have been pressed into use in attempts to describe them: thus there are batons, beans, boats, bowls, buttons, canoes, coins, crinkly-tops, cudgels, dumbbells, kidneys, lenses, pears, peanuts, spheroids, and teardrops, together with a whole host of other less descriptive comparisons. There is almost conclusive evidence that all the various shapes result from aerodynamic forces acting on a molten or partially molten form as the objects enter the high atmosphere of the Earth from outer space. The initial form of some of them at entry seems to have been spherical, and this shape is widely believed to result from surface tension when in the liquid form.

Table 6.2. Chemical composition of tektites.

	Average	Maximum	Minimum
SiO_2	73·40	82·68	68·00
Al_2O_3	12·70	16·46	9·56
FeO	4·37	6·81	1·13
CaO	2·40	3·92	0·04
MgO	2·28	4·05	1·15
K_2O	2·23	3·60	0·82
Na_2O	1·32	2·45	0·37
TiO_2	0·70	1·03	0·00
Fe_2O_3	0·41	2·25	0·00
H_2O	0·18	0·75	0·00
MnO	0·10	0·32	0·00

It is known from experiment that aerodynamic forces can contort plastic material into a great number of strange shapes.

Evidence of former spherical shape is in part preserved on the posterior surfaces of the button-type australites (Plate 7). Strong confirmation of the view that they have entered through the atmosphere is provided by the fact that it is now possible to manufacture these buttons artificially in a wind-tunnel and produce objects showing extraordinarily close resemblance in form to such australites. A sphere made of glass or from actual tektite material can be subjected to an airstream to simulate passage through the high atmosphere. It is found that, at speeds of the order of a few kilometres per second, heat developed in the boundary-layer in contact with the material is intense enough to cause this to melt and even to evaporate.

Upon heating, tektites are found to become deformable at about 800°C; complete fusion requires temperatures in the range of about 1100–1500°C for different tektites. These then are the temperature ranges needed for liquefaction to occur in the boundary-layer. When material of a tektite is converted to

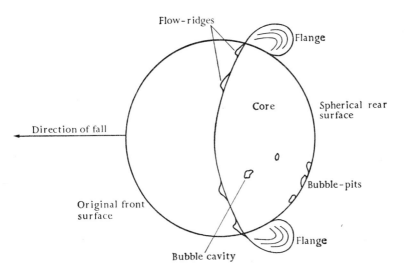

FIG. 6.1. Diagram illustrating how a sphere is changed to a button-shape by ablation. Some of the material removed from the anterior surface is lost altogether and some flows back to build up a solid flange. (The sketch represents a diametral section through the original spherical form.)

liquid or gaseous form it begins to take part in the surface flow created by the atmosphere streaming past. The boundary-layer forms on the anterior surface, and the flat-on orientation of the button-shaped tektites is in fact a stable aerodynamic position (Fig. 6.1). The heated liquid is swept along the surface rearwards until the boundary-layer itself begins to break away from the surface, which occurs about half-way round the (original) spherical form. This mechanism by which an aerodynamically heated object can lose matter from its surface is termed *ablation*. Some of the material may be lost altogether from the tektite, while the remainder, cooling as it moves towards the breakaway area, builds up to form a characteristic solid flange that transforms the shape to the remarkable button-like form of many australites.

Artificial tektites have been successfully made in recent years at a number of research centres, and the experiments indicate beyond doubt that speeds of low cosmic order are needed. There remains some disagreement on the precise figure, if indeed the same speed is involved for all tektites. The values arrived at range from about 6 km s^{-1} up to nearly 12 km s^{-1}. There is an inescapable difficulty in the experiments in that the glass spheres used have necessarily to be cool enough to be solid initially, whereas it could well be that freely falling natural tektites are still so much above laboratory temperatures as to be molten or plastic at the moment of entry into the high atmosphere, making conditions more suitable for ablation. This could, of course, be decided only if some reliable theory of their origin were available. Furthermore, there can at present be no guarantee that the Earth's atmosphere at these remote times in the past had exactly the same extent as now.

The button-like forms so common among australites are practically unknown elsewhere. This may result from the unlikelihood of incipient tektites entering the atmosphere with more or less perfectly symmetrical shape and in solid form. At speeds of the order of kilometres per second, aerodynamic pressures can far exceed the ambient atmospheric pressure, and will be different at different parts of the surface of a tektite. Their total effect may not reduce just to a simple force of resistance but could also involve twisting forces—couples—and these together could deform and contort a body having little internal strength. The degree to which a tektite had solidified at entry into the atmosphere might thus strongly influence the shape it developed in its fall. For many tektites, there are signs that ablation has taken place at most of the surface, and this suggests that for these stable aerodynamic orientation never took place, presumably owing to some irregularity of initial form.

The surfaces of tektites often show ridges, grooves, and waves resulting from flow during ablation; small pittings at the surface are also common. Some of these pittings may be produced by gas-bubbles bursting from just below the surface, but it has also been shown that pittings similar to those found can be produced by impacts of small particles. On certain bediasites, circular depressions of this kind are so numerous that their separating ridges are quite sharp. Surface features of this sort are absent only when there is other evidence of strong surface abrasion by weathering. Flow-patterns are found not only on the surfaces of tektites but almost everywhere within them, and especially within flanges when these are present.

Besides microscopic gas-pores, much larger bubble-craters also occur entirely within tektites; in some australites these range from 2 to 10 mm in diameter. These inclusions have the remarkable property that the gas contained in them is at very low pressure, usually less than 0·001 atm, despite the fact that the tektite may be millions of years old as measured by the time since its arrival on Earth. This feature may well provide one of the most important clues to the origin of tektites; it strongly suggests that they must have solidified in a locality where the surrounding gas-pressure was at very most of this order. In the terrestrial atmosphere, if no other forces were concerned, it would mean a height in excess of about 50 km. In order to balance the pressure resulting from surface tension on a small liquid tektite, the included gas would need to be at a pressure of only about 1 mm of mercury. The entire gas content of tektites is extremely small, 1 g of their material containing only of the order of 10^{-5} cm^3 by volume at normal temperature and pressure. A number of tektites have been found to shatter by implosion during laboratory handling, suggesting a permanent state of internal strain resulting from contained bubbles at less than atmospheric pressure.

The contained gases in tektites are almost entirely hydrogen, carbon monoxide, and carbon dioxide, with occasional minute traces of oxygen, nitrogen, and sometimes other gases. This markedly distinguishes tektite-glass from ordinary terrestrial glass of the obsidian type, since the gaseous content of the latter is found to be mainly chlorine and hydrochloric acid gas. Water is found in some tektites, but there is evidence that in these instances it has been absorbed after arrival on Earth. Fusion of tektites enables the rare gases, argon, neon, and helium, to be detected, though there is the consideration that tektite-glass may to some extent be permeable to helium.

The question of the ages of tektites is a complex and difficult one, not least because it is necessary to distinguish carefully exactly what is to be understood by the word 'age' in connexion with any particular line of argument. There is first the age of the material itself that goes to form the tektite. This would presumably mean the time since the stable isotopes within it were originally synthesized, supposing this to have occurred at the same time for them all. The ratio of the two isotopes of potassium, ^{39}K and ^{40}K, the second of which gradually changes into the first, would give this epoch, and it is found to be the same in tektites as it is for terrestrial potassium. This would seem to indicate that the material of the tektites originated at the same time as did that of the outer layers of the Earth; but as such an age probably holds for the other planets, and for the moon and asteroids, and possibly even for some regions of the galaxy, the coincidence provides little or no indication of the precise source of tektite material.

The principal ages associated with tektites, however, are clearly the times since their formation as individual bodies and the times since their actual final fall to Earth, whether through the high atmosphere, as certainly seems to hold for the australites, or as projectiles from remote terrestrial meteorite-craters, as some theorists consider possible. The difference between these times will clearly depend on the whole mechanism of formation; its determination must therefore be a theoretical question, and must remain so unless indeed we are fortunate enough to experience an actual fresh fall of tektites. The geological layers in which tektites are found are generally regarded as providing a good indication of the time of their fall, though this obviously relies on some element of assumption. However, on this basis the australites are easily the most recent with an age of something like 600 000 years, though some investigators put the age to be far less than this at only a few thousand years. The indochinites have similarly been associated with the mid-Pleiocene period and represent the next most recent fall. The moldavites are mid-Miocene; the javaites are mid-Pleistocene; the rizalites late Pliocene; and according to Barnes the bediasites, although now found in Pleistocene gravels, appear to have been derived from an Eocene formation, and this would make them the oldest of all tektites. These considerations do not, of course, serve to date the falls with any great accuracy, but it would seem fairly certain that all known ones have occurred within the last 50 million years.

There is direct evidence that tektites can have been entirely molten for a period of at most a few minutes. Under polarized light, tektites reveal series of dark and light bands associated with the internal stress-pattern; these bands run parallel to the intricate patterns of flow-layers, which in turn consist of material having slightly different compositions. This stress-pattern can be entirely removed by maintaining the tektite at a temperature in excess of about 1600°C for some thirty minutes, during which time the composition becomes quite uniform. This shows that solidification of tektites can have required a time of only a very few minutes. Rapid cooling from a temperature in excess of 1000°C could readily be explained by ordinary conduction within the tektite and radiation from its surface, for in the absence of any sources of heat a glass sphere of radius 1 cm would cool substantially (the excess of its temperature over its surroundings would fall by a factor of about 1/3) in about three minutes.

If the tektites formed and solidified in high vacuum long before reaching the neighbourhood of the Earth, it would seem that they would initially all have had spherical form (as a result of surface-tension forces). The button-shaped forms would then be expected to be of common occurrence, whereas in fact these perfectly symmetrical forms are to be found only among australites. This may be an indication that cooling and solidification occurred not long before, or even during, the passage through the Earth's atmosphere, at any rate for tektites other than australites.

Isotopic abundance-ratios in tektites are consistent with a terrestrial source of the material, but as yet there can be no certainty that they are inconsistent with an origin in lunar or cometary material, or indeed in other planetary material. And if meteorites are involved in the production of tektites, the composition of the meteorite concerned might well affect the final composition of the resulting tektites. Detailed chemical analyses by refined methods have tended if anything to obscure the problem of the origin of tektites, for, while some investigators insist that the results indicate a close resemblance to acid igneous rocks, others maintain there is similarity to sedimentary rocks, and yet others claim that they have no similarity to any rocks of terrestrial type.

The presence of minute traces of radioactive isotopes and the resulting production of argon are regarded as providing possible means of determining the ages of tektites. For example, the radio-active isotope of aluminium ^{26}Al is contained in australites in amounts indicating an age of less than half a million years, whereas there is no measurable activity of this element in either the moldavites or bediasites, implying for them an age not less than 4 million years. On the assumption that the argon content is purely radiogenic, upper limits to ages can be found, but these generally turn out to be far higher than the ages found by other methods. If the tektites have been exposed to cosmic rays, then certain unstable isotopes would be produced. Properly interpreted the amounts present must be some indication of age; but unless the full history of the material is known, any conclusions may depend very critically on the assumptions made, whether these are explicit or tacit. For example, it would be unsafe to suppose that the material of an actual tektite has always been combined in a single mass of its present size or thereabouts, for it is possible that it could have been part of a far larger body, and thereby have been mostly shielded from cosmic rays, or have existed as minute particles on which only primary cosmic rays would be effective. Thus any safe deduction of tektite ages will require a reliable theory of their origin, and this at present is the very thing that is not available.

Perhaps because they reach the Earth by high-velocity passage through the atmosphere, attempts have been made to classify tektites as a type of meteorite. But apart from this one feature in common, which is a purely dynamical consequence of gravitational effects, no serious comparison is possible. The extreme rarity of tektite falls and their restriction to but a few areas of the globe contrast strongly with the many hundreds of known areas of falls of meteoritic swarms; and although surface-ablation occurs also on meteorites, this is an inevitable consequence of their high speed in the atmosphere. Again, meteorite sizes and masses are for the most part vastly greater than those of tektites, and there is no resemblance whatever in their detailed compositions. Stony meteorites can be matched with terrestrial rocks, but there is no resemblance between tektite compositions and those of the terrestrial rocks of the areas in which they occur. Almost the only rock that tektites even partially resemble appears to be obsidian glass, a volcanic rock of high silica content. But even here the tektites have still higher silica content and a far lower water content; and the contained gases are quite different and much smaller in amount in tektites.

The origin of tektites has naturally aroused a great deal of speculation, ranging from mere suggestions taking little or no account of their properties to serious attempts at a scientific explanation supported by experiment and mathematical discussion. It would be a lengthy task to review all these many ideas, and we can perhaps omit from discussion such as those ascribing tektites to the activities of the aborigines in Australia, and the proposal, also made for comets, that they originate in solar prominences. Some of the suggestions have been concerned only with the observed distribution, as for example the 'great-circle hypothesis', which was advanced some forty years ago to try to account for all the then known fields by means of a single meteoritic episode. It was pointed out that a great-circle band on the Earth's surface a mere 20° wide, with its central line running from Tasmania in a roughly north-westerly direction to Czechoslovakia, would contain all these fields. A meteoritic swarm of small bodies was therefore postulated as having travelled over the Earth in this great-circle plane in such a way, not too clearly described in the theory, that the main proportion of them were deposited in and near Australia and the East Indies. Some preceding or following satellite-swarm was also postulated to account for the far-removed group of moldavites. The theory was really no more than a verbal description in which the things to be explained were simply restated in slightly different form as properties of some postulated but otherwise unknown agency, and it obviously gives no satisfactory account of the mode of formation or nature of individual tektites. It is little surprise that fields discovered since, or not taken account of in advancing the theory, lie far off even this broad band: the bediasites are some 30° off, while the Ivory Coast field is as much as 50° away. Moreover, the different ages of the fields are entirely against their having simultaneous origins.

A process of this type has recently been revived by O'Keefe in the proposal, which takes heart from observations of the Cyrillid meteorite shower of the year 1913, that a meteorite followed an orbit close enough to the Earth to enter the high atmosphere at its closest approach and underwent sufficient heating to shower off tektites on to an elongated area beneath its path. This mode of origin proposes that all the far-eastern fields are the result of a single such event. It seems a difficulty of the idea that known meteorites falling to Earth should according to this process also have produced something resembling tektites that ought to be found widely distributed over the surface. No such objects appear to have been recognized. The thickness of the boundary layer on a meteorite would depend mainly on the speed and viscosity rather than the size of the object, and such droplets as would be removed by ablation would probably be far below the size required for tektites.

The aerodynamic forms of australites make it possible to rule out any theory of an artificial origin, and there is little to be said for introducing any such hypothesis for the other fields, though the suggestion has been made that tektites may represent by-products of prehistoric glass-manufacture by man. Natural fires have also been invoked as a possible cause, but when the actual sites of recent fires are examined nothing remotely resembling tektites is found. The action of natural electric discharges in lightning-strikes has been yet another proposal, but rocks so fused and vitrified, which do in fact occur, are well known as fulgurites, and bear little or no resemblance to tektites except that they tend to a high silica content simply because this is the most refractory common constituent of rocks. In a similar way, nuclear explosions can also convert surface materials into glass, but there are no reports of any resulting tektites. Still another suggestion has been that tektites have formed as nodules within rocks, a process that is known to produce so-called geodes; but again these in no way resemble tektites. And there is the contrary suggestion that they have been formed from larger objects by prolonged abrasion. But the glassy structure of tektites strongly indicates a high-temperature process of formation lasting for only a brief period of time.

Most of these ideas have seemed to give so slight promise of successful explanation of tektites that little serious attention has been accorded them. But the same cannot be said of the volcanic hypothesis. Despite fatal defects, this has found strong appeal for many, just as has the volcanic theory of the lunar craters. Such an origin for the australites in particular dominated the subject during the nineteenth century to such an extent that it was inferred that there must at one time have been widespread volcanic activity on the Australian continent. But no direct signs of any volcanoes have yet been found, nor are there present any igneous rocks of composition similar to australites. Then again, where possibly suitable rocks are to be found, such as obsidian in New Guinea, there are no neighbouring tektite fields. The shapes also tell against a volcanic origin, for the actual ejections from recent volcanoes can be directly examined, and nothing remotely resembling tektite forms is found.

The possibility that meteoritic impacts on the Earth itself are the initial source of tektites has been proposed to parallel the volcanic hypothesis, but here again there is an absence of evidence of powerful meteoritic action associated locally with any of the known fields. For the older fields, with ages of the order of tens of millions of years, it could be argued, and indeed would have to be argued, that weathering and erosion had removed all traces of the original crater; but the argument might be difficult to sustain for the australite field, even if this is as much as half a million years old. On the other hand, if the area of the related meteoritic impact is not to be regarded as coinciding with the field itself, then the smallness of some of the fields, such as the moldavites and the Ivory Coast tektites, would seem to tell against the hypothesis, for it might be expected that the splashing-out of material, tektites included, would be more or less symmetrical even if it consisted of a number of fairly narrow radial streams, as the rays surrounding lunar craters might suggest. Then again, it simply is not known whether a great many other tektite fields, which would be expected to have been produced at the same time, lie hidden elsewhere on Earth, either buried on continental areas and heavily silted over or strewn in inaccessible ocean depths; so it is not possible to be absolutely certain that the distribution of tektites is entirely inconsistent with a terrestrial meteoritic origin.

A point in favour of the hypothesis is the discovery by Barnes that small glassy objects showing aerodynamic forms have actually been produced at the site of at least one meteorite crater, in Saudi Arabia. The objects concerned were, however, all found in the immediate neighbourhood. The difficulty with this form of the theory is to see how liquid droplets could be ejected with such violence as to have speeds measured in kilometres per second (in order to be projected to remote areas), and yet for the weak forces of surface-tension to take control of their shapes before they cooled. The australites would seem quite inexplicable on such a basis. The meteoritic explosion would have first to propel the incipient tektites from ground-level to somewhere outside the atmosphere; and so great would the air resistance and ablation be initially, in the dense lower atmosphere, that huge initial 'tektites' would be needed, and then surface-tension could scarcely operate. It is an interesting question what is the minimum size of object, solid or liquid, that could be ejected from ground-level to beyond the atmosphere with an initial velocity just adequate to achieve this. In launching a satellite, this part of the path is taken as slowly as possible with the express purpose of avoiding heating and possible ablation, and a second and subsequent stages are introduced only after most of the atmosphere has been left behind. On the other hand, supposing that an object could be so propelled, it could be argued that the very severe ablation would heat it as well as remove material from its surface; but whether the accompanying surface-heating could penetrate inwards sufficiently to liquify the interior, or maintain it liquid, so that surface-tension could eventually produce spherical forms, is yet another question that would need to be answered.

These objections, which have seemed to many to tell conclusively against a terrestrial origin, would not all necessarily apply if some similar external origin could be found. The nearest and most obvious source to be considered is, of course, the moon; and here to begin with the ever-to-hand volcanic hypothesis has again been pressed into service. The aerodynamic difficulties associated with the initial projection no longer arise because of the absence of any appreciable lunar atmosphere.

However, there is the severe difficulty that the principal lunar surface features are almost certainly not volcanic in origin at all, but the result of meteorite impacts; nor is it yet established that there has ever been any volcanic activity on the moon.

Accordingly, if the moon is the source of tektites, the mechanism of their removal must almost certainly have been meteoritic explosions. This hypothesis has been widely advocated in the past decade or so, especially for the australites, and until recent times there has been an almost complete absence of alternative theories. If it is possible for suitable liquid bodies to hold together in such an explosion, while at the same time being driven out at escape-speeds, then there would clearly be the possibility of some of them being intercepted by the Earth. The flight across the intervening quarter of a million miles would occupy a day or more, and this would give ample time for the complete solidification of the droplets, so that they would enter the high atmosphere with spherical form. This would be what is needed for the australites, at least for the button-like forms, but these particular shapes are practically absent from the remaining fields; this fact seems in some way to distinguish the australite field, and may perhaps be evidence against a lunar origin for at least some of the other fields.

Then again, if tektites are splashed off the moon in this way, it is difficult to see why they should land on the Earth in such small areas. As seen from the moon, the whole Earth presents a target occupying only 1/14 000th of the area of the sky, though for initial speeds near the moon of the order of 1 km s^{-1} the effective area of the Earth is increased about a hundredfold because of its strong gravitational field. Even so, the whole Earth remains a small target, while an area of a few thousand square kilometres represents a minute target indeed as seen from the moon—the whole area of Earth presented to the moon is about 250 million km^2. In order for a swarm to strike such a small target, the initial velocities would not only have to be very precisely directed but also have very accurately similar values, and it remains to be seen whether detailed dynamical calculations show signs of focusing effects that might be invoked to account for the limited areas of the known tektite fields.

A second consideration of importance, if not an outright difficulty with such a mode of origin, is that most meteoritic lunar impacts would simply splash off the supposed tektites into interplanetary space in directions not immediately bringing them to the Earth; but the subsequent paths of their orbits round the sun would obviously intersect or pass near that of the Earth, and at least an occasional tektite would fall like a meteorite. Such tektites should be found almost anywhere and might even be seen to arrive, but no such phenomenon has ever been reported. To meet this objection, it has been argued that a tektite moving freely in space would itself be subject to bombardment by tiny meteors and other particles that would etch away its surface, gradually reducing its size to meteoric dimensions. How valid this explanation may be cannot be finally settled as yet, since the dust-content of interplanetary space, and consequently its abrasive power, are not known with any certainty. For tektites produced as recently as the australites, it would require that reduction of radius by about a centimetre could proceed in a time of a million years. The recent photographs of the small-scale structure of the lunar surface taken by Luna-9 certainly suggest intense bombardment there by tiny particles, and if the lunar hypothesis of tektite origin is correct this evidence might suggest that surface-abrasion can effectively destroy them in interplanetary space in periods of time that are short compared with the age of the Earth.

The smallness of most of the tektite fields has seemed to many workers to provide conclusive evidence against a lunar origin. Urey in particular has drawn attention to the extreme difficulty experienced nowadays in the reverse process of landing a single space-probe on a selected area of the moon, even with the full operation of control-systems, in order to emphasize the unlikelihood of any natural process meeting the requirements for hundreds of thousands of projectiles. If, however, the moon is ruled out, where next can there be found a possible source?

Urey himself has come forward with the proposal that encounters of the Earth with comets may in some way or other be the prime cause. The mechanism as proposed by him is that a comet could enter the Earth's atmosphere and by a process of gaseous compression bring about heating effects at the surface, these being sufficiently intense to fuse locally the materials of both the Earth and comet, and in some manner thereby produce tektites. The details of this scorched-earth mechanism have not been made altogether clear even by their author, but there is reason to believe that the structure of a comet is entirely meteoric and that no effects of the kind envisaged could possibly result.

Nevertheless, encounters of the Earth with comets must certainly have taken place, and there is something to be said for investigating what might in fact then occur. If in its quiescent state a comet *were* a small object only a few kilometres in dimensions, as some astronomers maintain, then the encounter could be described as the comet meeting the Earth, but the frequency of such interactions would probably be inadequate, at something like once every 100 million years, to account for the number of tektite fields with their recent origins. Also the nature of the encounter would necessarily be akin to a meteorite of similar size striking the Earth's surface, and the resulting theory of tektite formation could not be importantly different from the meteoritic theory itself. On the other hand, as has been shown in Chapter 4, there is little or no basis for such a picture of cometary structure, and it is practically certain that comets consist of vast swarms of tiny *meteoric* (not meteoritic) particles spaced many metres apart and spread through an irregularly shaped volume that is commonly far larger than that of a planet, indeed sometimes comparable with that of the sun itself. In this case, it would be rather the Earth as a small body that encountered the comet, and the frequency of occurrence would now be about once every million years, a value of the right kind of order for association with tektite fields.

But the consequences of the encounter would now be quite

different, for we have to think of the tiny but massive Earth travelling through a very diffuse swarm of particles anything from ten to a hundred times as extensive linearly as the Earth itself. Particles of the comet that directly entered the atmosphere, as obviously would those that were met by the forward-moving hemisphere of the Earth, would simply behave as individual meteors, and although they would be very numerous none could penetrate to ground-level. It is not possible to suppose that objects sufficiently large to become tektites are contained originally within a comet, for then the whole surface would be strewn with them, contrary to the actual situation, and there is in any case indirect evidence that comets contain no large particles. This conclusion emerges from the evidence of meteor-showers, since however intense these may be no particles are ever observed to reach ground-level. In the great shower of the year 1833, it is on record that something like 200 000 meteors per hour could be seen from any single station in north America—the sky was described as like a snowstorm with them —and many thousands of millions must have entered the atmosphere during a time of the order of a day: yet there was no report of a single object falling to the ground.* And the same holds for all other meteor-showers, which are known to move only in cometary orbits.

The gravitational influence of the Earth upon the particles, which would otherwise simply stream by to the sides, would be to deflect them inwards towards the general axial line behind the Earth defined by the direction of the streaming. As in the accretion process of comet-formation, they will tend to be focused into a very narrow and much denser stream. The nearer such a particle passes to the Earth, the more it will be deflected, and the larger the angle and speed at which it enters the stream. The sideways motion will be destroyed, and the corresponding energy converted to energy of heat. Far out, deflection by the

---

* A comparably intense shower occurred in November 1966, and (despite some improvement in communications since 1833) again no report of any associated *meteoritic* fall was made.

Earth will be only slight, the particles will converge at only a small angle to the axial line, and little of the energy of motion will be destroyed (Fig. 6.2). Thus the inward pull of the Earth can control the inner part of the stream, and the same analysis that was used in relation to comets shows that, after a steady state has been reached, the inner part of the stream out to a certain distance actually flows inwards, which means vertically downwards on to the surface of the Earth. There is a neutral

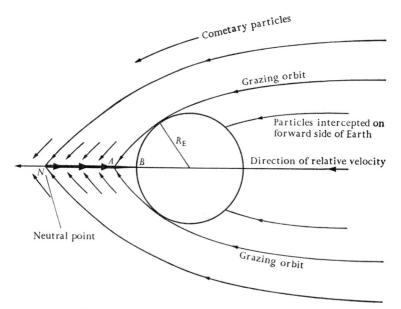

Fig. 6.2. Passage of the Earth through a comet.

point in the stream, and beyond this the flow is outwards and directly away from the Earth.

The heating produced by the energy brought in sideways would be sufficient to raise the temperature in the inward-flowing stream to over 1000°C, and in suitable circumstances this could cause the most refractory substances to melt. We thus have the possibility of a dense jet of extremely hot material being drawn down vertically to the Earth, and the proposal has been made that it is in such a jet that incipient tektites are first formed. Clearly the heating could drive off the most volatile constituents, and if not too intense it could leave the most refractory portion in liquid form.

Success of the process would evidently require that the neutral point should lie outside the solid body of the Earth; indeed, it must lie farther out than the point where those particles just grazing the high atmosphere at the sides converge together. This in turn means that the comet must not be moving too fast relative to the Earth, and it is of considerable interest that a proportion of short-period comets crossing the Earth's orbit would have sufficiently low speeds for this. The long-period comets are ruled out because they have come in from such great distances that they rush across the Earth's orbit so fast that their particles would only be encountered on the forward side of the Earth. It is not, of course, necessary that every encounter with a short-period comet results in the formation of such an inward jet, but the probability is about even judged by the present short-period comets. Calculation shows that the orbital eccentricity of a short-period comet would not have to exceed about 0·65, which would imply a relative velocity between the Earth and comet of about 8 km s^{-1}. All long-period comets have eccentricities of 0·99 or more, and their speeds are therefore too high for the formation of an accretion-stream by the Earth.

Although the interior part of the stream is falling towards the Earth, it is continually receiving outward thrust as a result of the particles joining it from the sides, though out to the neutral point, N (Fig. 6.2), this is not sufficient to overcome the inward attraction of the Earth. Fig. 6.2 shows that, closer to the Earth than the point A, the stream is shielded from receiving any further particles from the sides. Thus, for the section AB of the jet, the material is falling freely, and also is no longer receiving any supply of heat. The height of the point A above the effective surface of the Earth (which here, of course, would mean the level of the high atmosphere where meteors begin to be intercepted) is of the order of 0·1 to 0·3 times the Earth's radius, according to the relative speed of encounter; the time taken to fall freely through this distance would range from about two to seven minutes. This is just about the same order of time that it would take a sphere of glass of a centimetre or so in radius to cool substantially.

In this mechanism, the inflowing stream would probably consist of both molten materials, corresponding to the more refractory substances of the original cometary material, and gases corresponding to the more volatile components. The convergence of flow towards the general axial line implies that there would be an actual line or small realm of convergence where the transverse velocity was zero. Gases would tend to expand sideways and give the stream additional width, but the liquid portion of the stream might be quite narrow, and if the temperature produced by sideways-entering particles were sufficiently high the stream might be entirely gaseous at its outer parts. However, if the relative velocity of the Earth and comet were low, the point A where the supply of thermal energy ceases would be high above the Earth (Fig. 6.2), and the time of free fall from A to B, the high-atmosphere level, would be several minutes. In this time, cooling would produce liquefaction, and the most refractory substances would liquefy first. The downward stream would of course be accelerating, just as a stream of water from a tap accelerates under gravity, and surface-tension would take over in the stream to form droplets. These would be the initial tektites.

It is seen that the properties of such droplets would measure up to several of the requirements of tektites. If the liquefaction occurred in the accretion stream where the sideways supply of particles was still taking place, some of the incoming particles might well become embedded in the liquid while retaining some identity by remaining solid or partly so. The existence of such particles within some tektites is one of their most remarkable properties, as we have seen. Then again, if the time of free fall were long—it could be as much as seven or eight minutes—the liquid pre-entry tektites might even have time to solidify, which is what would be required for the button-shaped australites. Conversely, if the time of fall were much shorter, the tektites might still be liquid on entry, and aerodynamic forces could then distort them into the various forms that are found in other fields. Thus the relative velocity of the comet, which could be different on different occasions, provides a parameter within the mechanism allowing a variety of types of encounter to occur. The size of the comet or, what comes to the same thing, the length of path that the Earth describes through it, would also introduce another variable.

The place on Earth where tektites so formed would fall would depend on the direction of motion of the comet relative to the Earth, since the direction of the inward jet would be exactly opposite to this. The distribution of the planes of motion of short-period comets has its greatest concentration near the ecliptic, and it can readily be shown that as a consequence the places of fall are more likely to occur in low latitudes than in high, though no area of the Earth's surface is entirely inaccessible. In fact, with the exception of the Czechoslovak field at latitude 50°N, all the fields are between latitudes 39°N and 40°S.

At first sight this process might appear to direct all the tektites of any one fall to practically the same point of the Earth's surface, but a number of factors would tend to produce a much more widespread field. First, the accretion process is known to give a highly unstable kind of stream, and it may well tend to shower the droplets in a somewhat divergent way. Secondly, the aerodynamic pressures on the droplets when they reach the atmosphere may give considerable sideways forces leading to wide scatter at the ground. The tektites would have some hundreds of kilometres of atmosphere to traverse, and horizontal displacements by distances measured in hundreds of kilometres might well occur, since the sideways forces on an irregularly shaped object can easily be as large as or far greater than gravity. It is known that meteorites, for example, on entering the Earth's atmosphere can be disrupted by aerodynamic forces and the parts fall over a wide area. When other objects, such as aircraft, have broken up in flight, fragments have been scattered miles around, despite the fact that these were dense objects falling from only a few miles.

A third factor that must certainly extend the area of fall is plainly the rotation of the Earth. The time of passage through a comet would depend on its size and the line that the Earth happened to take through it, as well as upon the relative speed; it would be of the order of an hour, though clearly it would vary considerably according to the precise circumstances. In one hour, the Earth turns through about 1600 km at the equator; an east–west extension of this order could thus readily result. Through a combination of these three factors, the different dimensions of the known fields seem explicable, from the smallest at some tens of kilometres in linear extent to the largest at some thousands of kilometres.

The rotation of the Earth would, of course, extend the fields in an east–west direction, the eastern part of the field being the first to fall. There have in fact been some claims of evidence for systematic differences with longitude across fields, both in the australite field and the indochinites, the examples found in the eastern areas being systematically smaller than those to the west. However, where the oldest fields are concerned, the possibility has to be borne in mind that the axis of rotation of the Earth may not in the remote past have occupied the same position relative to the solid Earth that it does today. With the axis elsewhere at the time of a tektite fall, the extent produced by rotation could be in quite another direction than the present east–west system of parallels. This property of the mechanism at present might therefore be difficult to correlate with any but the most recent fields.

An important consideration that makes it difficult to judge whether this cometary mechanism can provide an adequate explanation of tektites is that so little is known of the composition and nature of cometary particles. An order-of-magnitude estimate of the total mass gathered in by the Earth during a passage through a comet can be made from analysis of the accretion process; the figure is something like a few thousand million tons. If only one part in a thousand of this gave rise to silica, that would nevertheless represent a few tens of millions of tons. Spread over an area of a million square kilometres this would provide some 10 g or so per square metre, which is very roughly of the observed order. But a cometary origin cannot at present be sufficiently well established to render unnecessary the search for some other process; indeed, there can be no guarantee that a unique process acting at different times has been responsible for the several fields.

A possible objection to a cometary origin is that the particles of the comet might probably have been long exposed to cosmic rays, and if so the tektites should contain a greater proportion of radioactive aluminium than they are found to have. This is one of the main reasons that a terrestrial source is favoured, but such an origin could be satisfactory only if the aerodynamic difficulties of projecting material from the surface right outside the Earth's atmosphere can be overcome. A conceivable method for doing this does in fact exist in the meteoritic hypothesis, provided that the explosion can be so powerful as to be able to blow off the whole atmosphere above and surrounding the area of impact. This may at first sight seem impossible, but the entire mass of the atmosphere is equivalent to a layer of only about 10 m of water and hence about 3 m of rock, whereas known large lunar impact craters are thousands of metres deep. Rare as sufficiently large meteoritic impacts might be, there is nothing to rule them out entirely. A mass of a few million million tons, which would mean a meteorite about 10 kilometres in diameter, striking the Earth at 20 km s^{-1}, would have energy of 2×10^{30} ergs, far and away greater than any bomb yet conceived, and a thousand times greater than the entire energy of the largest recorded earthquake. This is the size of meteorite that could have produced Copernicus or Tycho (a fairly recent crater judging by its intense rays); even larger meteorites would have been needed to produce the great lunar seas such as the Mare Imbrium.

An object this size, although its passage through the atmosphere would be an affair of incredible violence, would itself scarcely be resisted at all by the air, and would strike the solid ground at practically its interplanetary speed relative to the Earth. The impact speed would be so great that even the solid material could not be decelerated immediately, because the ordinary forces that give strength to materials are propagated too slowly to achieve this; instead, the materials of the meteorite and terrestrial surface coming into contact would intermingle to form a gas at density comparable to that of rock, corresponding to the density of the materials themselves, and at temperatures approaching a million degrees, corresponding to the original speed of motion. In such a gas, shock-waves can spread out at speeds comparable with but exceeding the speed of the incoming meteorite, and in this way the motion can finally be arrested. During a time of the order of a second, a meteorite 10 kilometres in diameter could be brought to rest after a penetration through the surface two or three times its own original length, and there would then be implanted several million million tons of gas at almost stellar temperature. It is the explosive expansion of this gas that produces the meteorite craters. But from the present point of view it is not the formation of the crater that is of interest, but the possibility that a large amount of this hot gaseous material may be ejected outwards to great distances in the form of a huge expanding cloud. The initial rate of expansion of the outer boundary of the gas would again be comparable with that of the impact speed, for it is this same energy that is the original cause, and so a range of radial speeds would be represented in the cloud. When gas expands in this way it automatically cools, and eventually it will reach such a degree of expansion that the temperature will fall locally to that at which liquefaction and even solidification of some of its constituents into small particles can occur.

The result of the explosion can thus be pictured as an ejection at all speeds up to something like 10 or 20 km s^{-1} of a radially expanding hemisphere of gas, some of which condenses into fine dust particles. Such an explosion finds least resistance vertically above the place where it occurs, and greater and greater resistance as the direction becomes more nearly horizontal, because there is a greater effective depth of material the nearer the motion is to the horizontal direction.

It cannot be suggested that tektites could form at once in such a holocaust, and the dust particles involved would be of extremely small size, such as occur in smoke. But where they were driven out into the empty space surrounding the Earth, their subsequent motions would have an effect rather similar to the accretion process for an encounter with a comet. Here, however, the cloud of particles emerges in all directions (above the horizontal), and would begin to describe orbits under the Earth's attraction. Those that initially move at less than 45° to the upward vertical would either escape altogether from the Earth or else fall back to the surface after describing less than 180° of angle as seen from the centre of the Earth (Fig. 6.3, p. 212). Those that fall back to Earth could therefore only have the effect of introducing a shower of dust into the high atmosphere, though now all parts of the Earth would receive such particles if the initial speeds were great enough. The speed required would be in excess of about 8 km s^{-1}, which is the speed that would just carry a particle round the Earth in a circular path.

Clearly a particle projected just above the horizontal with speed greater than 8 km s^{-1} would not reach the Earth's surface until more than halfway round, and so would pass through the axial line diametrically opposite to the point of impact of the meteorite (Fig. 6.3, p. 212). For greater elevation of projection, a sufficient speed would take the particles more than halfway round; a fountain of orbits will thus result, all converging to this axial line. We thus have a process closely resembling the cometary accretion mechanism, except that now there will be a range of initial speeds and initial angles of ejection (at less than 45° to the horizontal). In the same way as for the cometary mechanism, it can be shown that the converging paths of the particles lead to the formation of a jet, though now the particles all have an *inward* velocity component as they come to the axis. Thus an essentially downward stream results, with the velocities

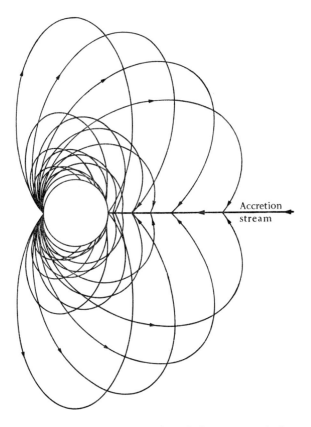

FIG. 6.3. Trajectories of particles ejected from an explosion at the surface of the Earth, such as might result from the impact of a meteorite. Only elliptical paths are shown.

of the particles such that considerable heating of the jet can again result; and accordingly tektites form in the molten part of the stream. Some slight asymmetry of distribution of the paths may be produced as a result of the rotation of the Earth, since this will in general add a small horizontal velocity component to the initial velocity of every particle. But as each particle starts out from the same point (the locality of the meteoritic impact), its path must later intersect the diameter (produced) of the Earth through the point of impact.

As far as the geometry goes, the downward jet in this mechanism would therefore be aimed at the point of the Earth's surface diametrically opposite to the meteorite impact point. The scene of fall of tektites produced in this way would thus be on the far side of the Earth from where the impact occurred. However, the time of flight of the various particles round to this opposite side would range from about one hour, for those of lowest speed and starting at a low angle, to a few or even several hours for those thrown far out on wide orbits. In the course of an hour the Earth turns 15°; so, instead of the tektites falling at the point diametrically opposite to the impact, the places of fall could be displaced by any values up to 90° to the west of this, and then spread some degrees in longitude, the particular amount depending on how long the downward jet persisted.

On this basis, for the australites an impact in the southern part of the United States would be suitably situated, and for the philippinites the corresponding region would be somewhere in South America. For the older fields, it has to be remembered not only that the axis of rotation may not have coincided with its present position, but that the configuration of the land-distribution could also have been somewhat different at the time of fall. No tektites have been found in the Soviet Union, which occupies one-sixth of the whole land area of the Earth, and it is of interest that this territory is mainly opposite to the Pacific Ocean. It is only recently becoming recognized that the Earth must have been just as frequently bombarded, area for area, as the moon, but there is the great difference that erosion and weathering of terrestrial craters would proceed with great rapidity and within a million years at most would remove almost all superficial trace of a large crater. Only a minute fraction of the area of the Earth has been drilled into by man, and most of this exploration has been carried out with other objectives than seeking for signs of meteoritic action, with the result that little is known about what present surface features may be the result of meteoritic impacts.

The essential element that differentiates this form of the meteoritic mechanism, and also the cometary mechanism, from the original form, is the heating process in the accretion stream. It would be this feature that would mainly determine the composition of the tektites, for the effect of cooling from a few thousand degrees would be that the most refractory substances would condense first to the liquid form, thereby giving surface-tension the opportunity to produce droplets, the volatile constituents meanwhile remaining gaseous and possibly escaping from the stream. A further interesting feature of this form of the meteoritic and cometary mechanisms is that either would explain why tektite compositions show no close correlation with those of the local rocks. It might be very worth while to examine rocks in regions more or less antipodal to the various fields with the object of seeing if any support for the hypothesis can thus be found.

The problem of the origin of tektites cannot be said to be anywhere near completely solved. Indeed, if the conclusions of all the many workers in the field were accepted, they would imply between them that tektites cannot possibly exist. But somehow they must have been formed, and their existence can only mean that at least some of the arguments and conclusions are unsound. Out of all the many studies that have been made there must surely be the possibility of getting to the truth, though it is always possible that some fundamental process has still been overlooked. The exhaustion of hypotheses is always difficult to achieve, and a new idea may arise at any time that might completely alter the picture. Meanwhile the tektites remain one of the outstanding puzzles of the solar system.

Uranus, but these could be calculated with high accuracy, and the observations adjusted accordingly. Had the solar system been effectively completed by the discovery of Uranus, and assuming the inverse-square law of gravitation along with the laws of dynamics to provide the correct theory, then the resulting adjusted orbit of Uranus should have been a Keplerian ellipse to within very small amounts that could well be attributed to observational error. But observations of the planet subsequent to 1781 soon began to show that this did not hold.

A number of suggested causes of a non-scientific kind were not long in forthcoming, as for instance the idea that at such great distances from the sun the law of attraction no longer followed the accurate inverse-square form. On the other hand, the hypothesis of a further planet superior to Uranus was an obvious possibility that must have occurred to many astronomers even before these irregularities of motion were established, but the steps needed to proceed from the bare idea to the actual discovery of yet another planet were apparently beyond the resources and courage of most of the astronomers of that day. Two men alone, Adams and LeVerrier, working quite independently and by rather different methods, carried through the mathematical labours that seemed to them to be involved, though the great German astronomer, Bessel, was devoting himself to the preliminaries for an attack on the problem in the months preceding his death on the 17 March 1846, just over six months before the date of the discovery of Neptune.

It is important for the purposes of the present discussion to emphasize at the outset that the success achieved by both Adams and LeVerrier was only partial. Though the discovery of the planet was manifestly achieved, there is considerable doubt as to the extent to which it was a logical consequence of their mathematical calculations. Each of them, however, had aimed not only at discovery, but also at what they believed to be essential for this—the calculation from the known residuals of Uranus of the whole set of orbital elements of the hypothetical planet together with its mass. It was fairly evident that if the correct explanation were an unknown planet it must lie farther out than Uranus itself, for the perturbations indicated a planet probably comparable with Uranus in mass, and this if closer in would be a whole magnitude brighter than Uranus and unlikely not to have been detected already. Moreover at an inferior distance inadmissible perturbations of Saturn might be introduced.

It was over this very question of the possible distance from the sun of the planet that Adams and LeVerrier were sadly misled. The fact that Uranus at 19.2 AU is just over twice the distance of Saturn from the sun, and that Saturn moves at approximately twice the distance of Jupiter, had for some time been associated into a more general empirical rule that had gained considerable currency under the somewhat unjustifiable title of Bode's *law*. Despite the fact that it is incorrect for Mercury, that it requires a major planet to lie between Mars and Jupiter, where there is not one, and fails dismally for Neptune, the 'law' seems still widely thought to have some special significance, and is even by some regarded as a desideratum of any theory of the origin of the solar system. The reason that Adams and LeVerrier relied on Bode's law, and thereby placed the unknown planet at approximately twice the distance of Uranus, was that, in the mathematical expressions for the forces produced by one planet on another, the ratio of distances a/a' (mean distance of Uranus/mean distance of Neptune) enters as an infinite series of powers that converge only slowly, and several such series are involved. In modern times these have all been fully tabulated to high accuracy for a whole range of values of a/a', and in most recent years they could be readily incorporated in a computer programme and evaluated for any required ratio. But no such facilities were available in the 1840s, with the result that Adams and LeVerrier almost of necessity succumbed to the temptation to rely on Bode's law. This meant that a large part of the calculations could be done once and for all, or at least carried out very few times, and then the remaining unknowns could be concentrated upon.

Both astronomers naturally assumed that Neptune moved in the same plane with Uranus, since this reduces the number of unknowns. The actual unknowns associated with the planet are then five in number: a', the mean distance from the sun (dealt with by means of Bode's law); e', the eccentricity; ϖ', the longitude of perihelion, which determines the orientation of the ellipse in space; τ', the instant at which the planet passes through perihelion; and m', the mass of Neptune. Before proceeding, we show in Table 7.1 the values actually found by Adams and LeVerrier for a', e', ϖ', and m' as compared with the values eventually found for the actual planet from observation.

Table 7.1. Values found by Adams and LeVerrier for Neptune compared with actual values

	Adams	LeVerrier	Actual
Mean distance from sun, a'	37·2	36·2	30·07
Eccentricity, e'	0·121	0·108	0·0086
Longitude of perihelion, ϖ'	299°	285°	44°
Sun/mass of Neptune	6670	9350	19 300
Longitude at discovery	330°·9	327°·4	328°·4

It will be seen at once from this Table that both Adams and LeVerrier evidently had some reservations about the precise validity of Bode's law. Indeed, Adams had regarded an earlier solution adopting $a' = 38\cdot4$ AU exactly as less satisfactory, and presumably LeVerrier had similarly concluded that the value given by Bode's law was too large. However, at that time the idea of mathematical proof and of scientifically established principles had scarcely even begun to catch on, and with the strong influence of mumbo-jumbo descriptions heavy upon mankind, a predilection existed for claims such as that of Bode's law. It can further be seen from Table 7.1 that in both cases e' came out extremely large by most planetary standards, and this might itself have been grounds for suspicion even before discovery, while none of the quantities a', ϖ', and m' really agreed well with the actual values subsequently found from observations of Neptune itself.

Nevertheless the planet was found within about 1° (twice the diameter of the moon) of the place calculated by LeVerrier, and within about $2\frac{1}{2}$° of the position found by Adams. The whole area of the sky is more than 40 000 square degrees, and even if it were supposed that the planet would be within a few degrees of the ecliptic, say ± 5°, that still would mean an area of nearly 4000 square degrees. So LeVerrier's prediction improved the chance of finding the planet by a factor of about a thousand.

Before describing the circumstances of the actual discovery it is necessary to say something of the earlier history of the matter, and in particular the means sought by Adams to bring his work to the attention of the authorities and those having at their disposal telescopes that might be capable of detecting the planet. This also brings us to the personal side of the matter and to the sources of the great controversy that followed upon the discovery, which was acclaimed far and wide as the greatest triumph of mathematical science ever achieved.

To take Adams's role in this drama first: he was still an undergraduate at Cambridge when, as early as the summer of 1841, he conceived the notion that the irregularities of Uranus, then exciting the astronomical world, were due to the action of an exterior planet. And in early 1843, soon after he had taken his degree, which he did with such distinction as should have warranted him more consideration thereafter from his seniors, he settled to the numerical work that eventually led him to his prediction, as given in Table 7.1.

Before the end of the year 1843, Adams had made a preliminary solution that was sufficiently promising (in its power to reduce the unexplained displacements of Uranus in the sky) to convince him that his hypothesis must almost certainly be correct. At this stage, Adams desired to have the most up-to-date observational material concerning Uranus, and through the influence of Challis, then the professor of astronomy at Cambridge, the Greenwich observations were supplied by the Astronomer Royal of the day, George Airy, himself formerly a

Cambridge mathematician, and before 1835 the holder of the chair currently held by Challis and also director of the observatory there. Not only that, Airy like Adams had been Senior Wrangler in his turn, a mathematical distinction that in those days immediately marked a man as of the highest potentialities. Airy certainly possessed great ability, but he seems soon to have forgotten that others of equal or even greater capacity might follow on his heels and have views worthy of every respect. However, by September 1845, some two years later, Adams had completed his calculations and had arrived at a solution that seemed to account fully for the motion of Uranus, and the problem then arose of what he was to do about it.

In those days, scientific societies and scientific journals were few, and such means of communication as existed were slow and inadequate. Indeed, not long before this Adams had actually published his results for the elements of the periodic comet Vico–Swift in *The Times* (15 October 1844)! To meet the situation, the plan was therefore adopted of communicating the results to the Astronomer Royal, and for this purpose Challis provided Adams with a letter of introduction to Airy. The letter, which was to be delivered personally by Adams, outlined the position, but left the results to be explained by Adams himself. Unfortunately, Airy was away in France and therefore Adams did not meet him, but upon his return he evidently heard about the visit of Adams to Greenwich and wrote to Challis expressing a desire to learn more of the investigations by correspondence. This was in late September 1845, and four weeks later, on the way back to Cambridge from his home in Cornwall, Adams again made his way to the Royal Observatory. Once again, unhappily, Airy was not at home; and when Adams returned an hour later he was informed that the Astronomer Royal was at dinner, but there was no message from Airy as to what course the young Adams might follow in order to see him. Adams, however, did manage to leave a statement of the results of his calculations showing how the residuals of Uranus were well accounted for, together with a note containing the predicted elements of the planet, its mass, and its mean longitude on a stated date. From this information, the predicted position on any other date could readily be found by an elementary calculation.

All this happened fully six months before any other announcement was made anywhere else in the world respecting the problem. By some manner of means Adams had surmounted all the difficulties and had arrived at a location only a couple of degrees or so from where the planet actually was in the sky. All that remained in order to obtain for Adams and for English science the undivided glory of the achievement was a careful search of the heavens in the neighbourhood indicated, a search that could not have failed, in a few weeks at most, to reveal the planet. But the necessary comparatively minor effort was not forthcoming on the part of the observational astronomers that knew of the matter.

Two weeks went by before Airy replied to Adams's note by a letter in which he opined that the perturbations produced in Uranus by a planet with 'certain assumed elements' seemed extremely satisfactory. Adams of course had not simply 'assumed certain elements'; he had actually arrived at these as the end-product of a calculation, and it must have been very noticeable to him, after spending literally years on the calculations of these very elements, that the Astronomer Royal above all persons had apparently not even grasped what he was trying to do. It must be remembered that Adams was still at an age when he would assume that a person in high position was thoroughly in touch with every aspect of the subject that his appointment required him to be conversant with. But however competent a person may be, it is still necessary in addition that he keep his hand in, whereas just at this period Airy was much occupied with the non-astronomical problems of the Railway-gauge Commission. In his letter, Airy went on to ask the question whether the assumed perturbations would also explain the errors of the radius-vector of Uranus. Here again Adams must have been most irritated by this. The work had concentrated on the errors in longitude, which can be determined with far higher accuracy than any of the errors in distance. Several years before, in 1838, Airy himself had shown that there were discrepancies in the radius-vector of Uranus, and evidently this loomed large in his mind. But any such discrepancies must have been unimportant compared with the mounting errors in longitude.

At all events Adams did not feel moved to reply to the letter, thereby incurring the displeasure of the tribal chief of British astronomy, a thing that he may have regretted later on. Clearly Airy had not quite understood what Adams was attempting in his work, and may even have imagined that Adams had simply guessed at all the elements, as indeed he had done for a'. Although it is generally agreed that Airy acted somewhat regrettably at this stage, and even more so later on, it is necesary to bear in mind the kind of difficulty that may well have presented itself to him, difficulties of such a kind that only an experienced dynamical astronomer might appreciate, and yet perhaps too subtle or vaguely apprehended to be set down clearly. For example, Airy himself had discovered a new large inequality (perturbation) in the motion of the Earth and Venus that had been overlooked by earlier investigators. It happens that eight times the mean angular rate of Venus round the sun is nearly equal to thirteen times that of the Earth, indeed so very closely so that the difference is only about 1/240th of the Earth's mean motion, and hence producing a period of about 240 years. In integrating the equations of motion to determine the longitude, what would be expected to be a very small term is in fact enhanced by a factor of about two million, and there results a large perturbation of long period. A similar large inequality occurs more patently in the theory of Jupiter and Saturn: here five periods of Jupiter are closely equal to two of Saturn, so that small divisors resulting from integration arise much earlier in this case in the successive approximations to the path. This particular perturbation takes 913 years to run through all its values, and through it the effect on Jupiter is a term with amplitude about 28 minutes of arc, while in Saturn it is as much as 48 minutes, the difference being to some extent accounted for by the different masses.

It is sometimes naïvely thought that the perturbations of one planet on another must be greatest when the two planets are closest together, that is, near conjunction; but this is by no means the case. The mutual forces may be greatest then, but the resulting perturbations are two integrations (or even more) removed in highly non-linear differential equations, and no safe reasoning can be conducted other than by formal analytical processes. In fact, for the principal perturbations always present between two planets, the greatest perturbations in longitude tend to occur fairly near opposition, when the planets are almost farthest apart.

But the presence of particularly close accidental resonance can lead to unusually large perturbations, as the foregoing examples show. Accordingly Airy would have been especially alive to this possibility, in view of his own experience with the Earth–Venus problem, and it may well have perplexed him very much how observations of Uranus over a mere fifty to sixty years could lead to a determination of all the major sources of perturbations between Uranus and another hypothetical planet. It would have been by no means clear, and is still not clear today, how it could be shown *a priori* that discovery of such a planet is possible from a very limited range of observations. On the other hand, it equally cannot be claimed that no means of discovery can be found. Airy may well have reasoned that he could not see how it could possibly be done, and hence that it could not be done, but evidently this last conclusion does not necessarily follow, as he was to find to his chagrin only too soon.

If Adams had been the only person to work out a solution to the problem and predict a position, Airy's lack of enthusiasm might simply have produced delay in the ultimate discovery, but in fact the matter was destined to be complicated and brought to the level of an international scandal by the

circumstance that the leading French dynamical astronomer
LeVerrier had also been systematically attacking the problem.
After first re-discussing all existing observations of Uranus,
LeVerrier pronounced it certain that known actions were
incapable of explaining the observations, and he also came to
the conclusion, as Adams had done before him, that the only
hypothesis capable of meeting the situation was that of an
external undetected planet. LeVerrier then set about the
formidable task of finding an orbit and mass for such a planet,
but made use of a somewhat different method from that of
Adams. On 1 July 1846 the results of his calculations first saw
the light of day in the *Comptes Rendus*, and a copy of the
actual publication reached Airy a few weeks later, now a good
seven months after his receipt of Adams's prediction. Airy
curiously enough noticed at once that the position arrived at
for the planet by LeVerrier was only just over 3° away from
that given by Adams. This above everything else did much to
lessen Airy's misgivings about the whole business, and, as he
later wrote, he felt no doubts as to the accuracy of both
calculations, as applied to the perturbations of longitude.

But Airy still had misgivings about the radius-vector, and he
thereupon wrote to LeVerrier putting this query again, though
he appears not to have taken the opportunity to inform
LeVerrier of the parallel investigation in England. LeVerrier,
besides being nearly ten years older than Adams, was a far
more worldly and redoubtable character, and he did not make
the mistake of failing to reply to a high dignitary such as Airy.
Indeed, he wrote at length explaining exactly how his theory
must automatically take account of the discrepancies in radius-
vector, and he also cleverly utilized the opportunity to express
the hope that Airy might arrange for a search to be made for
the unknown planet, offering to send further information in the
form of more precise positions for the planet at future dates.
Airy's burst of enthusiasm was short-lived, however, and he
made no reply to this offer by LeVerrier. On the other hand,
quite soon after this, there took place the annual meeting

of the formal governing body of the Royal Observatory. Anyone that has attended such meetings will appreciate how thankful Airy must have been to have the subject of what he termed the 'new planet' available to give himself something of an astronomical nature to speak about, though it can obviously have had little relevance to the official business of the Board. Airy went so far as to present it as an example of the benefits of international co-operation, and to explain that there was now an extreme probability of a new planet being discovered very shortly, and he referred to the close agreement between the results of Adams and LeVerrier.

Fired evidently by his own oratory, Airy at last began to feel that something must be done to search for the planet. The time was now July 1846, and he accordingly wrote to Challis at Cambridge, where the Northumberland telescope was particularly suited to the task. Airy suggested that Challis should search a band of the heavens measuring 30° along the ecliptic and 10° wide centred close to the positions indicated by Adams and LeVerrier. It happened that no star-chart of this area had yet been published by any observatory, so Challis determined to cover the zone three times at separate intervals and map all stars down to magnitude 11. The planet itself, from the computed mass, was expected to be of about magnitude 9 and to show a disk about 3 seconds in angular diameter. The project would have involved recording so many individual stars that some 300 hours of observing time would have been required, and bearing in mind time lost through adverse conditions this would have meant several months' work at the telescope. Challis quite clearly entertained no great expectation of finding any planet, but nevertheless was willing to be guided by Airy in the matter. However, in a period of about two months beginning late in July and covering August and almost all of September 1846, Challis and his assistants actually measured the positions of over 3000 stars. Challis's lack of enthusiasm and generally negative attitude led him on inexorably to the eventual disaster: the planet was actually observed twice in only the first four days of his search, and had he compared the stars observed on 12 August with those of 30 July, he would have found that the forty-ninth of these was missing from the earlier field. This was in fact Neptune, since carried into the field of view by its planetary motion. Again, on 29 September, Challis saw a star that he actually noted down 'seems to have a disc', but despite this strong indication of its non-stellar character no effort was made to check it against earlier measures. This again was Neptune, but news of its discovery had not yet reached Challis.

While this funereal pace was being maintained at Cambridge, the energetic LeVerrier was busy exploring ways and means of instituting a search for the planet, but still quite oblivious of Adams's work and the search being made by Challis. In France, it seems, the same sort of reluctance to rely on 'theory' prevailed as in England, and the astute LeVerrier soon withdrew any reliance on his unenthusiastic compatriots. At this point LeVerrier decided to write to Galle at the Berlin Observatory. This step was favoured by two pieces of good fortune, though Encke himself, the director of the observatory, when he learnt of LeVerrier's theories was by no means enthusiastic about them. The first stroke of good luck was that Galle was still young enough to feel enthusiasm for the proposal and that on the staff there was also a second younger astronomer, named d'Arrest. Together they were prepared to persuade the reluctant Encke to let them undertake a search. Second and luckier still, they happened to have the star-map of the vital zone available to them, for these charts were actually prepared at Berlin and the relevant one was just about to be generally published. (Challis had ascertained before embarking on the construction of such a map that the all-important chart had not yet been received in Cambridge.)

Armed in this way, the discovery was simplicity itself for these skilful young astronomers, and the planet was duly located on the very first night, 23 September 1846, within only half an hour of the search beginning, when an object of magnitude 8 was observed that was not present on the chart. Neptune was discovered. The actual position was just under 1° from the longitude predicted by LeVerrier and just over 2° from that computed by Adams. The object was checked again the next night, and the motion meanwhile exhibited settled its planetary character beyond all doubt. Accordingly Encke announced the discovery publicly the following day, 25 September 1846. The news very understandably aroused tremendous enthusiasm in scientific circles everywhere, and indeed throughout the world, though it took time to travel and was not known in England until 1 October.

At first, naturally enough, it was LeVerrier's name alone that was associated with the theoretical prediction, but when the news did reach England an extremely difficult and delicate situation was created, and the dilly-dallyings of Airy and Challis began to come under serious scrutiny and criticism. The story of all that then followed has been told and mistold in varying degrees on many occasions since, and we need not here again go into the devious claims and excuses that were put forward. Suffice it to say that it soon became clear that as far as the reputation of British science was concerned a superlative opportunity had been culpably thrown away. The advancement of astronomy itself was held up scarcely a year, and manifestly the discovery had been made entirely without the efforts of Adams, Airy, and Challis playing any part, though in fairness there is the reservation to be made that Airy had made it clear to LeVerrier that little help could be expected from him, and he thereby deflected the Frenchman to seek elsewhere for practical aid to the discovery.

On the discovery becoming generally known, minds were made up with little or no appreciation of what was really involved, excesses of adulation were launched in many directions, and lofty phrases, such as 'this magnificent triumph of astronomical theory', 'the greatest mathematical achievement of all time', 'remarkable calculations resolving the problem of inverse perturbations', were much to the fore. The successful outcome, in the matter of actually finding the planet, naturally gripped everyone's imagination and quite justifiably created great excitement. Moreover this very success tended to divert all attention from the interesting technical questions posed by the remarkable fact that the actual orbit of the newly discovered planet, very soon afterwards computed directly from observation, differed so much from the theoretically calculated ones. In fact, all that could certainly have been claimed was that a planet had been found near a place that emerged from a long series of theoretical procedures. But the exact status of the discovery was by no means clear: it may be that all the steps then made can be shown to have complete validity, but this has not yet been done even to this day.

In describing the work as representing the very zenith of mathematical skill, there was perhaps a pardonable element of exaggeration, bearing in mind the atmosphere of great excitement that prevailed in 1846. But even in the centenary celebrations of 1946 there was a considerable amount of repetition of the lavish claims of a century before. All praise is due to both Adams and LeVerrier, of course, but from a scientific point of view it has long since become necessary to attempt to form a proper assessment of their work. There can be no doubt of their clarity of mind and scientific correctness in perceiving that the hypothesis of another planet was the one to be examined to the utmost limits before considering a change in the law of gravitation, or suchlike, and there need be no reservation either about their assiduousness in carrying out the vast amount of calculation that their methods inevitably involved. But the mathematical side of the problem was not really of great difficulty or complexity to anyone with competent acquaintance with dynamical astronomy, and the analytical expressions that LeVerrier, for example, utilized were of an inevitable kind that already existed in planetary theory, except that it was naturally necessary to use symbols to denote elements and other quantities of unknown value. As we shall see, LeVerrier at the outset was faced with determining as many as nine unknowns, and with about a century and a half of observations of Uranus available, a truly vast amount of arithmetic was needed to take adequate account of all these in order to arrive finally at 'best' values for the elements and mass of the unseen planet.

In the event, LeVerrier guessed at one of the unknowns, the distance of Neptune, by using Bode's law as we have seen, but the adopted value implied a period of the unknown planet of some 218 years, which is not very different from three times the period of Uranus, namely 84 years. It would be sufficiently near in fact to give a mild resonance, because of the approximate 3:1 ratio. In Adams's calculation, with $a' = 37\cdot2$, the corresponding period would have been 227 years, and so would lead to even closer 3:1 resonance. But in fact the period of Neptune finally turned out to be 165 years, which is a very close 2:1 resonance with Uranus. In the analytical forms for the perturbations, the terms associated with these two resonances have quite different coefficients and of course different periods, and accordingly, because of their reliance on Bode's law, both LeVerrier and Adams were using numerical coefficients for the various terms expressing the perturbations in longitude that gave quite wrong emphasis to the respective terms for the 2:1 and 3:1 resonances. It is still not clear even today how and why their calculations survived this.

A second obscurity in the work of Adams also remains unexplained to this day, for at the outset of his procedure he declared the existence of an advantage in working in terms of the perturbations in mean longitude of Uranus rather than with those of its true longitude (the latter being those found most directly from observation). The reason for this preference in Adams's own words was as follows:

'It is easily seen that the series expressing the corrections of *mean* longitude in terms of the corrections applied to the elements of the orbit is more convergent than that which gives the corrections of *true* longitude, and the same thing is true for the perturbations of the mean longitude, as compared with the true. The corrections found were accordingly converted into corrections of mean longitude by multiplying each of them by r^2/ab, r being the radius vector, and a,b the semi-axes of the orbit of Uranus.'

Now whereas such a relation as that claimed subsists *in an entirely undisturbed path* between the rates of change of true longitude v and of mean longitude $l = nt + \varepsilon$ in virtue of the angular momentum integral

$$r^2 \frac{dv}{dt} = h = nab,$$

which may be written

$$(r^2/ab)\, dv = ndt = dl,$$

there seems no validity in supposing for perturbed motion that the unexplained variation of true longitude Δv is similarly related to that of mean longitude Δl. No such relation seems known in celestial mechanics, and indeed if it existed the conversion made by Adams could only have been equivalent to multiplying both sides of an equation of condition by the same factor, namely r^2/ab, whose values presumably Adams calculated (approximately) from the available 'best' orbit of Uranus.

In the hope of discovering justification for this procedure, the writer has consulted several dynamical astronomers of the present day, but without success; and study of the literature reveals that no less a person than E. W. Brown, for many years regarded as the highest authority in all matters of dynamical astronomy until his death in 1939, remained puzzled by this step. Available accounts of Adams's treatment, as for example that given in his *Collected Works*, are not adequately detailed, and E. W. Brown could perforce only remark after his kindly way that 'the meaning of this process is not clear', and express the possibility, or perhaps hope, that 'the change may have been taken care of some other way'. In fact, when applied to the residuals, it means for the most part multiplication by a factor not very different from unity (the maximum value of r^2/ab for Uranus is about 1·18), with the result that the general trend of residuals with time is not much changed, and probably the actual mean-longitude residuals would show a similar form.

It has been conjectured that this curious step may account for Adams's final predicted place, $2\frac{1}{2}°$ away from the actual planet, being rather poorer than that of LeVerrier at a shade under 1° in longitude. But this explanation is extremely doubtful, for had the search been delayed only a few years Adams's prediction would have been distinctly improved while the orbit found by LeVerrier would have given a position moving away from the actual planet. Thus in the year 1856, for example, the longitude predicted by the orbit of Adams would have coincided almost exactly with that of the planet, while that from the orbit of LeVerrier would have been 3° away. Nor is it possible to gainsay this by invoking the influence of later information, if such were included, for as we have seen in the neighbourhood of conjunction perturbations of longitude tend to be smallest; conjunction actually occurred late in the year 1821, and, with the synodic period of 171·4 years that Uranus and Neptune have, a mere ten years of additional observations near conjunction would not mean moving into a realm for which the perturbations were unusually important.

Let us come now to the essentials of LeVerrier's method, which is more suited to our purpose than that of Adams since in itself it is free from any obscurity. The standard elliptic orbit of a planet is that obtained by first removing the action of all gravitational influences other than that of the sun, and so long as all sources of perturbations can be included a unique ellipse can be arrived at, apart from the minute irremovable effects assigned to errors of observations. Any attempt to find such an orbit for Uranus was therefore bound to fail, and any set of values of a, e, ε, ϖ, for this planet would need to have unknown corrections δa, δe, $\delta \varepsilon$, $\delta \varpi$ added to them. These then were *four* unknowns to be found. Since the values for Uranus were already known approximately, these corrections would occur only linearly in the expressions for the errors in longitude. In addition there were the entirely unknown a', e', ε', and ϖ', associated with the orbit of Neptune, together with its unknown mass m', making apparently *nine* in all. It happens that although ϖ' occurs trigonometrically, e' and ϖ' together can be replaced, in the expression for the errors in longitude, by $h' = e' \cos \varpi'$

and $k' = e' \sin \varpi'$; h' and k' are then found to occur only linearly. The mass m' multiplies everything throughout, so that m', $m'h'$, $m'k'$, etc., can be used as variables, and creates no difficulty, but a' and ε', in the complete absence of any known approximations to their values, can in no way be replaced by variables occurring linearly. Accordingly, as aforesaid, a' was assigned by Bode's law, but ε', which occurs as $\sin \varepsilon'$, $\cos \varepsilon'$, $\sin 2\varepsilon'$, $\cos 2\varepsilon'$, etc., might have any value from $0°$ to $360°$, and LeVerrier adopted the plan, with the aim of eventually arriving at linear equations, of taking in turn forty different values for ε' spaced at $9°$-intervals covering the whole possible range from $0°$–$360°$, and for each such value of ε' solved, by the method of least-squares, eighteen consolidated equations of condition based on observations of Uranus over the whole period 1690–1845. These eighteen equations would therefore be in *seven* unknowns. To solve a single such set of equations by ordinary long-hand methods of arithmetic would be a lengthy and tedious task, but that he carried it out forty times gives some indication of LeVerrier's determination. From these forty different solutions, it was then possible to select that one giving the closest fit to the whole series of observations. In this way LeVerrier arrived at $\varepsilon' = 252°$ as the initial best value, and then once such an approximately satisfactory value was available it could be slightly adjusted and improved by further use of linearized equations, now in *eight* variables. This in outline was LeVerrier's method, and it is not difficult to realize that a very large amount of arithmetic must have been involved. The same was true for Adams's method, and he too laboured intensively at the purely numerical side of the problem long after its theoretical nature had been decided upon.

But there remains the highly interesting question whether it could have been possible to discover the planet by more direct methods not involving such extensive calculations. That this might be so seems first to have been conjectured and discussed by E. W. Brown in 1931, who came to the conclusion that for this purpose the important thing to discover first from the observations would be the time of conjunction, and that if this could once be done little more might be needed to find the planet. But some doubts can be felt about the method then proposed by Brown for, quite apart from the fact that he was not always the most lucid of expositors himself, it gives the result that conjunction occurred close to the year 1840, which in fact means almost 90° away from the actual position, and would have led to a predicted longitude for Neptune some 27° ahead of its actual place at the time of discovery, even assuming a circular orbit for the planet at exactly the right distance. So Brown's method can scarcely be regarded as having any great power. The problem was taken up again by J. E. Littlewood and the present author following the 1946 centenary celebrations, and we agreed that the modern form of the problem could be put as follows: 'What is the simplest theoretical approach and minimum amount of calculation that could have led to the discovery of the planet?' This to some extent reflects the attitude of the pure mathematician when it is felt that some result established by laborious means must be capable of simple proof. But there remains the difference that the reasons for the success of the methods used by LeVerrier and Adams have still to be made clear even to this day, and until this is done there must remain a strong suspicion that they were favoured by a great deal of good luck, possibly in studying the problem at a time when their methods chanced to work.

The principal quantity to be compared with the theoretical disturbance produced by the unknown planet is the longitude of Uranus as seen from the sun, but as all observations are made from Earth the unexplained residuals have first to be arrived at by calculation. These reductions are of a formal character and need not detain us, and we can regard the basic material as the difference in heliocentric longitude as between that found from observations and that given by the ellipse of closest fit for the period prior to 1840 after the removal of the effects of Jupiter and Saturn. This material, as tabulated by Adams, is shown in Table 7.2. LeVerrier, in his discussion, included the

the result that h would be increasing; the converse would apply after conjunction. But near conjunction, the rate of change of h can be expected to be small, since Neptune will be pulling more or less in the radial direction. This means that a curve of h against t would be flat at and near conjunction, and therefore the precise point at which the rate of change vanished would be poorly determined unless h were extremely accurately known over the crucial range. In practice, this possibility fails altogether, however, because although v is well determined the

Table 7.3. Smoothed values of $\delta(t)''$ (seconds of arc)

Year	$\delta(t)('')$	Year	$\delta(t)('')$	Year	$\delta(t)('')$
1780	3·46	1796	21·45	1825	18·16
1781	5·0	1797	21·8	1826	17·0
1782	7·0	1798	22·0	1827	14·6
1783	8·45	1799	22·25	1828	10·82
1784	9·8	1800	22·5	1829	7·8
1785	11·5	1801		1830	2·0
1786	12·80	to 22·75	1831	− 3·98
1787	14·1	1816		1832	− 9·6
1788	15·25	1817	22·76	1833	−15·2
1789	16·40	1818	22·78	1834	−20·80
1790	17·45	1819	22·80	1835	−28·08
1791	18·25	1820	22·4	1836	−35·36
1792	19·00	1821	20·87	1837	−42·66
1793	19·9	1822	20·97	1838	−50·66
1794	20·6	1823	20·3	1839	−58·65
1795	21·0	1824	19·6	1840	−66·64

radius r cannot be found with anything approaching the same accuracy, and the resulting plot of $r^2\dot{v}$ against time produces a curve far too irregular, as a result of random errors, to permit the point of vanishing gradient to be read off with anything like sufficient precision.

But it happens that a far more accurate method based only on manipulation of the longitude v of Uranus is possible. To explain this: suppose we have a planet m moving in an accurate elliptic orbit of small eccentricity e, (for Uranus $e = 0\cdot047$), so that anything depending on e^2 or higher powers can be neglected.

Then its longitude, as seen from the sun, at any time t is given by the well-known expression.

$$v = nt + \varepsilon + 2e \sin (nt + \varepsilon - \varpi), \qquad (7.1)$$

where $2\pi/n$ is the period of the planet (84 years for Uranus), ϖ is the longitude of the perihelion point, and ε is the mean longitude when $t = 0$. As we have seen, the value of n (which is related to a by $n^2a^3 = 1$, in suitable units), and of e, ε, and ϖ available were necessarily in error by small amounts depending on the mass m' of Neptune, which is such that $m' \ll 1$ ($=$ mass of sun). Accordingly, as calculated at any time from slightly erroneous elements, v would itself be in error by

$$\begin{aligned}
\Delta v = {}& t\Delta n + \Delta\varepsilon + 2\Delta e \sin (nt + \varepsilon - \varpi) \\
& + 2e(\Delta\varepsilon - \Delta\varpi) \cos (nt + \varepsilon - \varpi) \\
& + 2et\,\Delta n \cos (nt + \varepsilon - \varpi) \qquad (7.2)
\end{aligned}$$

In this, the planetary equations show that each of the first four terms on the right must be of order m', but the last term is of order em', since Δn is of order m', and thus is so much smaller than the other four that it can be neglected. It is this property of Δv that enables the time of conjunction to be found with considerable accuracy by elementary and quite brief methods, as it is next proposed to show.

The above expression, omitting the last term, can be rewritten in the form

$$\Delta v = m' (p + qt + c \cos nt + d \sin nt), \qquad (7.3)$$

where $m'p = \Delta\varepsilon$, $m'q = \Delta n$, and $m'c$, $m'd$ are simply related to other fixed quantities in (7.2). Suppose now we denote by E the elliptic orbit best fitting the observations of Uranus, m, *after* the effects of all known planets have been removed. This would simply be the very ellipse available to Adams and LeVerrier that soon failed to satisfy the future observations with sufficient accuracy. Then we can write

$$\left.\begin{aligned}
v_E(t) &= \text{longitude at time } t \text{ as calculated from } E, \\
v(t) &= \text{actual observed longitude of Uranus.}
\end{aligned}\right\}(7.4)$$

Then the residuals in longitude (the quantities tabulated in Tables 7.2 and 7.3) for any given time are

$$\delta(t) = v(t) - v_E(t). \tag{7.5}$$

Suppose next that $t = t_0$ is the *unknown* instant of conjunction. If at that instant the action of Neptune were imagined to cease, then Uranus would move on in a ellipse E_0, say. This would be the so-called 'instantaneous ellipse' corresponding to the instant t_0; it is of course not known, but necessarily would differ from E only by small amounts proportional to m', since if $m' = 0$ the two ellipses would coincide. Let us write

$$v_0(t) = \text{longitude of Uranus in ellipse } E_0. \tag{7.6}$$

Next let $w(t)$ be the difference between the true longitude of Uranus and the longitude that the planet would have at time t in the instantaneous ellipse E_0, so that

$$w(t) = v(t) - v_0(t). \tag{7.7}$$

Then $w(t)$ vanishes for $t = t_0$, since by definition the actual orbit and the instantaneous orbit coincide at $t = t_0$. Hence we have from (7.5) that

$$\delta(t) = [v_0(t) - v_E(t)] + w(t). \tag{7.8}$$

In this last relation, both $v_0(t) - v_E(t)$ and $w(t)$ are small and of order m', as we have seen, and smaller terms of order em' are ignored. So in calculating $w(t)$ for present purposes, any term involving the eccentricity can be ignored, and this means in effect that both Uranus and Neptune may be regarded as moving in circular orbits. It is here, as a direct consequence of this, that the important point emerges that the values of $w(t)$ are accordingly equal and opposite at equal times before and after conjunction, for with circular orbits everything is symmetrical about the line sun–Uranus–Neptune at conjunction, and the motion is completely reversible with time about this. The precise analytical form of the function $w(t)$ is not required at this stage, and all that is necessary is to note that, if τ denotes time measured from conjunction, then clearly

$$w(t_0 - \tau) = -w(t_0 + \tau). \tag{7.9}$$

Or, if we write

$$W(\tau) = w(t) = w(t_0 + \tau), \tag{7.10}$$

then the relevant property of W that can in fact be used to find the unknown t_0 is that W is an *odd function* of τ; thus

$$W(-\tau) = -W(\tau). \tag{7.11}$$

Returning now to equation (7.8) for $\delta(t)$, the difference $v_0(t) - v_E(t)$ represents the difference in longitude between two elliptic orbits whose elements differ only slightly by amounts of order m', and this renders the form (7.3) equally applicable here. Thus (7.8) becomes

$$\delta(t) = m'\{p + qt_0 + q\tau + c\cos(nt_0 + n\tau) + d\sin(nt_0 + n\tau)\}$$
$$+ W(\tau), \tag{7.12}$$

and, by introducing different equivalent constants, this can be written finally as

$$\delta(t_0 + \tau) = A + B(1 - \cos n\tau) + \{C\tau + D\sin n\tau + W(\tau)\}, \tag{7.13}$$

and here now that part of $\delta(t)$ in the curled brackets is an *odd function* of τ. It is this feature of (7.13) that enables the unknown t_0 to be found. For this purpose we accordingly have

$$\delta(t_0 - \tau) + \delta(t_0 + \tau) = 2A + 2B(1 - \cos n\tau), \tag{7.14}$$

and since $\delta(t_0) = A$, we finally obtain

$$\rho(t) \equiv \frac{\delta(t_0 + \tau) - 2\delta(t_0) + \delta(t_0 - \tau)}{1 - \cos n\tau} = 2B = \text{constant.} \tag{7.15}$$

The numerator here is a second difference in the residuals derived from observation, and may be written

$$\Delta^2(\tau, t_0) \equiv \delta(t_0 + \tau) - 2\delta(t_0) + \delta(t_0 - \tau), \tag{7.16}$$

while in the denominator n is the mean angular rate of Uranus. We therefore have a simple criterion for finding t_0 from the available data that may be stated as follows: Select a series of instants t_0 for conjunction, and for each t_0 calculate $\Delta^2(\tau)$ by use of Table 7.3 and divide by the factor $1 - \cos n\tau$; then the resulting quantity $\rho(\tau)$ will be constant if t_0 has been selected at conjunction.

The constancy of $\rho(\tau)$ cannot of course be expected to be exact, because of observational errors and the small terms of order em' neglected in arriving at the result, but from a series of adopted values of t_0 the most nearly constant set of values of $\rho(\tau)$ can readily be found. Table 7.4 shows the results of such calculations with t_0 taken at three-year intervals from 1816 to 1828, and with τ starting at six years and increasing by steps of three years. It is plain that, for small values of τ, $\rho(\tau)$ would tend to be very unsteady, both because random errors would more strongly affect $\Delta^2(\tau)$ and because $1 - \cos n\tau$ is small. This is borne out by the values in Table 7.4 for $\tau = 6$. For this same reason, in computing the mean-square differences of $\rho(\tau)$ for each t_0, increased weight was attached with increasing τ in proportion to $\tau - 3$.

Table 7.4. Values of $\rho(\tau)$ at five selected instants for t_0

	$t_0 = 1816$	$t_0 = 1819$	$t_0 = 1822$	$t_0 = 1825$	$t_0 = 1828$
τ	$\rho(\tau)$	$\rho(\tau)$	$\rho(\tau)$	$\rho(\tau)$	$\rho(\tau)$
6	9·0	23·7	42·3	88·4	108·4
9	10·5	27·6	53·1	78·8	95·1
12	15·8	35·6	53·1	74·7	87·0
15	23·6	38·5	54·6	70·9	—
18	28·5	42·1	55·2	—	—
21	33·6	45·1	—	—	—
24	38·1	—	—	—	—
m.s.d.	10·0	6·8	3·5	6·1	9·4

Note: The figures for 1816 can range 24 years on either side because the observational material of Table 7.2 extends to 1840, but for 1819 the range is reduced to 21 years, and so on, with the result that there is one fewer entry in succeeding columns.

It is seen at once from Table 7.4, both from the individual values and from the mean-square differences given in the last line, that $\rho(\tau)$ is most nearly constant for 1822. By graphing the values of the mean-square differences (and using a somewhat greater range of t_0), it has been found that the best value of t_0 on the data of Table 7.2 is at 1822·3.

In this comparatively simple way we therefore arrive at the important result:

$$\text{Time of heliocentric conjunction} = 1822\text{·}3 \qquad (7.17)$$

As a matter of comparison, it may be noted at this point that the actual instant was in fact a little less than six months earlier at 1821·74. The adoption of 1822·3 in fact places the unknown Neptune just over 1° ahead of its actual longitude at this time.

In carrying the matter further it is convenient to adopt a time of conjunction coinciding with the exact beginning of a year, but here the obvious rounding-off to 1822·0 would bring us more than halfway nearer to actual conjunction. Accordingly to avoid the possibility of introducing any helpful improvement in this way, it was decided to adopt 1823·0 as the calculated instant of conjunction, and proceed on this basis.

To make a preliminary assessment of how important a knowledge of conjunction may be, it is a simple matter to calculate for any assumed size of circular orbit for Neptune where the planet would be 23·73 years later, that is at the actual time of its discovery 1846·73. The angular rate of such a planet is proportional to $a'^{-3/2}$, where a' is its assumed distance, and the curve of Fig. 7.1 shows the resulting position in longitude for values of a'/a ranging from about 1·4 to 2·1 ($a = $ mean distance of Uranus). The curve shows at once that a range in longitude of $\pm 10°$ on either side of the actual discovery position would be covered by the relatively great range of a'/a from about 1·43 to 1·89. Even on the assumption of Bode's law, $a'/a = 2$, the predicted longitude is only about 13° behind the actual position—accurate enough therefore to put it within the zone of longitude suggested by Airy for the search.

Had Bode's law in fact held fairly accurately, the present method as it stands would have given the position of the planet (using 1822·3 for conjunction) within about 1° of the true position. But in view of the serious failure on this unique occasion of the law, the question that arises is whether by further simple methods, again not involving arduous arithmetic, a close

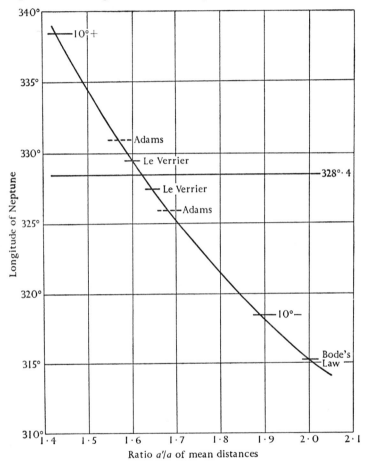

FIG. 7.1. Curve showing the predicted positions of Neptune at the date of discovery (1846·73), based on conjunction at 1823·0, for values of the distance ratio a'/a ranging from 1·4 to 2·1. The long horizontal line shows the actual longitude of Neptune at discovery (328°.4), and the short lines the permissible limits of a'/a for errors not exceeding those made by Adams, by LeVerrier, for an error within $\pm 10°$ of longitude, and for Bode's law.

estimate of the mean distance could be found. It appears in fact that this is to some extent possible, but it is unavoidable that any such method must now contemplate other values of a/a', and the numerical summing of the infinite series giving the values of the perturbations must be tackled. A certain amount of arithmetic is therefore inevitable, but the series concerned are fairly rapidly convergent, and with results for only a few values of a/a', interpolation can be used for intermediate values, as will be seen, if a certain regrouping is introduced.

With only hand-calculation available, the natural objective of any process would be to arrive at linearized equations for the unknowns, so that these could be solved by least-squares. As far as the corrections to the 'best' orbit for Uranus are concerned, these enter only linearly as we have seen, but for the others the form taken by the principal perturbations in longitude of m due to m' is given by a series of terms of the form

$$P(t) = m' \sum_{1}^{\infty} F_i \sin i \{(n' - n)t + \varepsilon' - \varepsilon\}$$

$$+ m'e \sum_{-\infty}^{\infty} G_i \sin [i \{(n' - n)t + \varepsilon' - \varepsilon\} + nt + \varepsilon - \varpi]$$

$$+ m'e' \sum_{-\infty}^{\infty} H_i \sin [i\{(n' - n)t + \varepsilon' - \varepsilon\} + nt + \varepsilon - \varpi']$$

$$(7.18)$$

to the first order in m', and in the eccentricities e and e'. In this expression, the coefficients F_i, G_i, and H_i are functions of the ratio of the mean distance, a/a', and have elaborate forms. It is these very quantities that one would wish to avoid computing over and over again if a' were itself regarded as an unknown. For instance,

$$F_i = \frac{i(z_i^2 + 3)}{z_i^2(1 - z_i^2)} a A^{(i)} + \frac{2z_i}{z_i^2(1 - z_i^2)} a^2 \frac{\partial A^{(i)}}{\partial a}, \qquad (7.19)$$

in which $z_i = i(n - n')/n$, with i taking successive integer values, and where

$$A^{(i)} = \frac{2}{a'} \cdot \frac{1 \cdot 3 \cdot 5 \ldots (2i-1)}{2 \cdot 4 \ldots \ldots 2i} \left(\frac{a}{a'}\right)^i \left[1 + \frac{1}{2} \cdot \frac{2i+1}{2i+2} \left(\frac{a}{a'}\right)^2 \right.$$

$$\left. + \frac{1 \cdot 3}{2 \cdot 4} \cdot \frac{2i+1 \cdot 2i+3}{2i+2 \cdot 2i+4} \left(\frac{a}{a'}\right)^4 + \ldots \right], \qquad (7.20)$$

with somewhat similar expressions for G_i and H_i. These show that, unless a definite numerical value is assumed for a/a', there is no possibility of obtaining completely linear equations, and this plainly is why both Adams and LeVerrier were tempted to adopt Bode's law to overcome the difficulty.

Now the smallness of the eccentricities of the planetary orbits could equally be said to be a 'law', and it turns out that considerable progress can be made if we adopt $e' = 0$ in place of $a/a' = 1/2$. At the price of some arithmetic in computing a few of the F_i's, this assumption immediately removes two of the unknowns associated with the Neptune orbit, namely e' and ϖ', and incidently all the terms of (7.18) in $m'e'$ vanish, which otherwise might require the computation of some of the H_i. Also in (7.18), since $e = 0 \cdot 047$, the terms in $m'e$ are much smaller than those in m' alone, and accordingly (though this would not be essential to the method to be described) it is proposed to calculate the perturbations as if Uranus itself were moving in a circular orbit. This reduces the computations and in the end leads to predictions of accuracy comparable to those of Adams and LeVerrier.

The complete expression for the residuals in longitude of m will then consist of two parts: first will be that depending solely on the corrections required to the original best-fitting ellipse; this takes the form

$$\Delta v = \Delta \varepsilon + t \Delta n + 2 \sin M . \Delta e - 2 \cos M . e \Delta \varpi, \quad (7.21)$$

where M is the mean anomaly $nt + \varepsilon - \varpi$. This expression is unaltered in form whatever instant t is measured from, so that it is always permissible to write

$$\Delta v = \Delta \varepsilon + t \Delta n + \Delta \alpha \sin nt + \Delta \beta \cos nt, \qquad (7.22)$$

where $\Delta \alpha$, $\Delta \beta$ are linear combinations of the original unknown corrections to the elements and are therefore themselves still small. Accordingly, if forthwith t is regarded as measured from conjunction, which can now of course be regarded as known with all-sufficient accuracy, the full expression for the residuals in heliocentric longitude, the quantities of Tables 7.1 and 7.2 in fact, will be given by

$$\Delta v = \Delta \varepsilon + \Delta n + \Delta \alpha \sin nt + \Delta \beta \cos nt + P(t), \quad (7.23)$$

where now the perturbation terms $P(t)$ reduce simply to

$$P(t) = -m' \sum_{i=1}^{\infty} F_i \sin i(n - n')t = -m' \sum_i^{\infty} F_i \sin iD, \quad (7.24)$$

where $D = (n - n')t$.

If a direct comparison were made between (7.23) and the observational material (Table 7.1 or 7.2), with an assumed value for a', and hence n' related to it by $n'^2 a'^3 = 1 = n^2 a^3$, to give a set of equations of condition, the unknowns would be five in number, namely $\Delta \varepsilon$, Δn, $\Delta \alpha$, $\Delta \beta$, and m'. This is already a substantial reduction on the number involved in the methods of Adams and LeVerrier, but it would still be necessary to try a series of values of a' in order to get the best fit. In view of this, and of a special other extremely important consequence, it is convenient to introduce the device of a certain simple regrouping of the equations of condition that has the effect of eliminating $\Delta \alpha$ and $\Delta \beta$ altogether.

To see how this can be done, consider the trigonometrical identities

$$\begin{aligned} \sin (x + \lambda) - \sin x + \sin (x - \lambda) &= (2 \cos \lambda - 1) \sin x \\ \cos (x + \lambda) - \cos x + \cos (x - \lambda) &= (2 \cos \lambda - 1) \cos x \end{aligned} \quad (7.25)$$

If in these $\lambda = 60°$, the righthand sides obviously both vanish. Now for the orbit of Uranus, the period $2\pi/n = 84 \cdot 013$ years, and accordingly 60° of mean longitude is described in almost exactly 14 years. Hence, if instead of a particular residual $\Delta v(t)$, we combine them three at a time and use

$$\Delta v(t + 14) - \Delta v(t) + \Delta v(t - 14), \quad (7.26)$$

the time t now being measured in years from conjunction, then the terms in $\Delta\alpha$ and $\Delta\beta$ in the corresponding theoretical expression resulting from (7.23) will be automatically eliminated. This regrouping of the equations of condition therefore results in a further reduction of the number of unknowns at the price of reducing the number of equations of condition, since t cannot now be taken less than 14 years from each end of the range, and of increasing the mean-square error associated with each equation of condition by a factor $\sqrt{3}$, since there are three observational terms combined now in (7.26). But these effects are more than offset, as will be seen next, by the complete elimination of the awkward effect of resonance near $n/n' = 2$, which as we have seen must otherwise enter the actual Neptune problem strongly.

Under this regrouping, the terms $\Delta\varepsilon$ and Δn are left unaltered but the terms of $P(t)$ have to be adjusted. We have in fact

$$P(t) = m' \ (F_1 \sin D + F_2 \sin 2D + F_3 \sin 3D + \cdots), \quad (7.27)$$

where $D = (n - n')t$, and it turns out that for a/a' in the range of interest here the series of coefficients $F_1, F_2, F_3, \ldots$ diminish fairly quickly with increasing suffix, with the exception of F_2 near the resonance $n = 2n'$ where, as (7.19) shows through the factor $1 - z_i^2$ in the denominator, it can become very large. Accordingly, where the new equations of condition are concerned, the appropriate value of λ involved in the identities (7.25), since the corresponding $x = (n - n')t$ for the term in $\sin D$, and $2(n - n')t$ for the term in $\sin 2D$, and so on, will be given by

$$D + \lambda = (n - n')(t + 14)$$
$$= D + \left(1 - \frac{n'}{n}\right)60°, \quad (7.28)$$

since $n \,.\, 14$ years $= 60°$. Writing $\nu = n'/n$ for the ratio of mean motions, $\lambda = (1 - \nu)60°$. Accordingly to form

$$Q(t) \equiv P(t + 14) - P(t) + P(t - 14), \quad (7.29)$$

the coefficients F_i in (27) must be multiplied in turn by

$$\left.\begin{array}{ll} 2\cos\{(1-\nu)60°\} & -1 \quad \text{for sin } D \\ 2\cos\{(1-\nu)120°\} & -1 \quad \text{for sin } 2D \\ 2\cos\{(1-\nu)180°\} & -1 \quad \text{for sin } 3D \end{array}\right\} \qquad (7.30)$$

and so on, where $\nu = n'/n$. We can write the resulting series for $Q(t)$ formally as

$$Q(t) = -m'(f_1 \sin D + f_2 \sin 2D + f_3 \sin 3D + \ldots), \quad (7.31)$$

and the revised equations of condition take the form

$$\Delta\varepsilon + t\Delta n + Q(t) = \Delta v(t + 14) - \Delta v(t) + \Delta v(t - 14). \quad (7.32)$$

Now, for given a/a', only *three* unknowns are involved, namely $\Delta\varepsilon$, Δn, and the mass m' which occurs in $Q(t)$.

Actual application of the method requires first the formation from Table 7.2 of the quantity $\Delta v(t + 14) - \Delta v(t) + \Delta v(t - 14)$, and the values of this at three-year intervals are found to be as in Table 7.5.

Table 7.5. *Values of* $\Delta v(t + 14) - \Delta v(t) + \Delta v(t - 14)$

Year	($''$)	Year	($''$)
1796	8·30	1814	10·57
1799	12·00	1817	− 3·99
1802	15·25	1820	−20·45
1805	18·30	1823	−40·21 (conjunction)
1808	18·82	1826	−60·89
1811	17·21		

The next requirement is the calculation of the coefficients in $Q(t)$. To begin with, this requires the calculation of the coefficients of the terms in $P(t)$ for a set of values of a/a' selected to cover any expected value of a'. These are shown in Table 7.6.

Table 7.6. *Coefficients of terms in* $P(t) = m'F_i \sin iD$

a/a'	sin D	sin $2D$	sin $3D$	sin $4D$
0·50	− 5·74	1·43	0·14	0·03
0·55	− 7·43	3·38	0·30	0·07
0·60	− 9·91	15·71	0·67	0·15
0·65	−13·72	−38·42	1·69	0·37
0·70	−19·94	−18·66	5·46	0·97

As Table 7.6 shows, the coefficient F_2 of $\sin 2D$ becomes large as the resonance $n/n' = 2$, or $a/a' = 0{\cdot}630$, is approached from either side. To avoid excessive calculations of the various F_i for different values of a/a', it would obviously be convenient if we could interpolate within Table 7.6; but in the case of F_2 interpolation with much accuracy near the resonance would clearly not be possible from just a few values. It happens that the Uranus–Neptune system, with $a/a' = 0{\cdot}638$, comes much too close to this resonance for interpolation of Table 7.6 to be at all possible for this specially important coefficient.

It is here that the transformation to $Q(t)$ has special value, for the conversion factor $(2 \cos \lambda - 1)$ of (7.25) vanishes precisely at the resonance where the coefficient F_2 has infinite value through the factor $1 - z_i = 2\nu - 1$ in its denominator. The expression

$$\frac{2 \cos \{(1 - \nu)120^\circ\} - 1}{2\nu - 1}, \qquad (7.33)$$

which occurs in f_2, has as its limit $\pi/\sqrt{3}$ as $\nu = n'/n \to \frac{1}{2}$, and accordingly the coefficient f_2 passes smoothly through the resonance without singularity, as Table 7.7 of the transformed coefficients now shows:

Table 7.7. Coefficients of terms in $Q(t) = - m'\Sigma f_i \sin iD$

a/a'	$\sin D$	$\sin 2D$	$\sin 3D$	$\sin 4D$
0·50	3·21	0·81	0·27	0·09
0·55	4·66	1·29	0·47	0·18
0·60	6·88	2·05	0·82	0·35
0·65	10·32	3·30	1·44	0·67
0·70	16·17	5·47	2·56	1·28

The equations of condition (7.32) can now be solved for a series of values a/a' with the secure circumstance that all the numerical coefficients $f_1, f_2, f_3, \ldots$ change quite smoothly as a/a' changes. The resulting set can be tested for each value of a/a', by a least-squares method, and it is found that there is a very pronounced 'best-fit' at just less than $a/a' = 1{\cdot}6$, which corresponds to a mean distance of 30·71 AU, as compared with

17—M.S.S.

the true value of 30·07. The present method could of course be applied in far greater numerical detail, but there is clearly little point in doing more than merely demonstrate the practicality of the procedure.

Consistently with the greater proximity of a' to the true value, the present method also gives a closer value for the mass of the planet than was found by either Adams or LeVerrier (Table 7.1), though still rather strangely discordant. The value arrived at is in fact

$$m/\text{sun} = 1/25{,}500. \tag{7.34}$$

The longitude predicted for the actual date of discovery is readily found to be 329°·4 compared with the true value of 328°·4. The error of the present method is therefore about 1° ahead of the right position, whereas that of LeVerrier was just under 1° behind.

It is a simple matter to find the longitude that the solutions found by Adams, and LeVerrier, and by the present method would predict for instants other than that of discovery. The three curves of Fig. 7.2 show the differences between the actual longitude of Neptune and the longitudes calculated by the three methods. It can be seen at a glance from these that whereas LeVerrier's result has always been regarded as superior to that of Adams, in fact a few years later quite the reverse would have held, supposing the same observational material to have been utilized. On the other hand, the curves show that an accuracy within $\pm 1°$ would have obtained on Adams's solution for only about five years, and on LeVerrier's solution for about twelve years, while on the method described here assuming a circular orbit this interval is about twenty-four years. These are, of course, consequences almost entirely of the degree of closeness with which the quantity a' is found.

It is important to remember that both Adams and LeVerrier intended their solutions to lead to an orbit accounting for the residuals over the whole range of observations utilized. The observations used by Adams covered the range from 1712 to 1840, while those used by LeVerrier were from 1690 to 1845, and in fitting an analytical expression to such a range there is no reason to expect that the resulting orbit will be especially accurate, for prediction of Neptune, at one time rather than at another, though it might be expected on several grounds that it would fit best near the middle of the range, which would mean

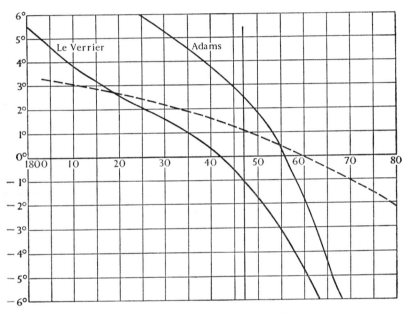

FIG. 7.2. Curves showing the amounts by which the orbits of Adams, LeVerrier, and the circular-orbit solution (dashed curve) differ from the true position of Neptune. (The vertical line, at 1846·73, shows the date of discovery.)

near 1770, rather than at one end that chanced to be the important one. But the predicted position for Neptune for the date 1770 would have been nearly 30° wrong on the basis of the Adams orbit and not much less than this on that of LeVerrier. The actual time of prediction of interest was, of course, at the extreme end of the whole range, and fortunately it was with this that the two orbits agreed, as Fig. 7.2 shows, and not the other end. Good luck indeed for Adams and LeVerrier, but not so good for Airy! This does not apply, however, to the method described above based on first finding the instant of conjunction. That the difference of longitude as shown by the dashed curve increases (in magnitude) in both directions is simply due to the slightly inaccurate value of a', which could probably be improved by more detailed arithmetic.

To conclude the present account of this outstanding episode, it is worth recalling that the passage from data to conclusions is one of the chief objectives of science. But it is also something that we are having to do at most moments of our lives. We seldom notice how readily we effect this passage and how much assumption is involved, especially when strong emotion is present. Then we make the step so easily, and often so assuredly, that we fail to perceive that an intricate subjective psychological process must be concerned, of a non-deductive character, and one that almost defies analysis. But it is a main principle of scientific method that every step should be brought to the surface, as it were, and its validity demonstrated, since until this has been satisfactorily accomplished a matter cannot be regarded as properly understood, and any conclusions otherwise arrived at may not be valid. As we have seen in the present chapter, this has yet to be carried out even for the Neptune problem; and so the true meaning and status of the discovery of Neptune—all glory to Adams and LeVerrier nevertheless being allowed—must still remain an open question.

Selected Bibliography

THE ORIGIN OF THE SOLAR SYSTEM

JEFFREYS, H. *The Earth*, 1st edn. 1924; 2nd edn. 1929 (Cambridge University Press).

RUSSELL, H. N. *The solar system and its origin* (Macmillan, 1935).

JEANS, J. H. *Astronomy and cosmogony* (Cambridge University Press, 1929).

HOYLE, F. *Frontiers of astronomy* (Heinemann, 1958).

JASTROW, R. and CAMERON, A. G. W. (editors). *The origin of the solar system* (Academic Press, 1963).

SCHMIDT, O. *The origin of the Earth* (Lawrence and Wishart, 1959).

THE INTERIOR OF THE EARTH

JEFFREYS, H. *The Earth*, 4th edn. (Cambridge University Press, 1959).

———, *Earthquakes and mountains* (Methuen, 1950).

BULLARD, E. C. and others. *Advances in geophysics*, Vol. 3 (Academic Press, 1956).

HOLMES, A. *The age of the Earth* (Nelson, 1957).

BULLEN, K. E. *Theory of seismology* (Cambridge University Press, 1964).

KUIPER, G. P. (editor). *The Earth as a planet. The solar system*, Vol. 2 (Chicago University Press, 1954). [This comprehensive work contains large numbers of references on many aspects of geophysics and related problems.]

CONSTITUTION OF THE TERRESTRIAL PLANETS

JEFFREYS, H. *The Earth*, 4th edn. (Cambridge University Press. 1959), Ch. 4.

UREY, H. C. *The planets* (Yale University Press, 1952).

VAUCOULEURS, G. DE. *Physics of the planet Mars* (Faber and Faber, 1954).

RUSSELL, H. N. *The solar system*. Vol. 1 of *Astronomy* (Ginn, 1945).

KUIPER, G. P. *and* MIDDLEHURST, B. M. *Planets and satellites. The solar system*, Vol. 3 (Chicago University Press, 1961). [This volume consists of eighteen separate contributions to planetary physics and the dynamics of the solar system, and each appends a large number of up-to-date references.]

THE NATURE OF COMETS

RICHTER, N. B. *The nature of comets* (Methuen, 1963). [Translated and brought up to date to September 1962 by Dr Arthur Beer. This work contains practically complete lists of references to papers on comets up to 1962.]

WHIPPLE, F. L. *The Earth, moon, and planets* (Harvard University Press, 1963).

PORTER, J. G. *Comets and meteor-streams* (Chapman and Hall, 1952).

WURM, K. *The physics of comets: the solar system*. Vol. 4 (Middlehurst, B. M. and Kuiper, G. P. editors), (Chicago University Press, 1963), Ch. 17. [This volume contains a number of articles on special cometary hypotheses together with numerous references.]

LYTTLETON, R. A. *The comets and their origin* (Cambridge University Press, 1953), Chs. 1 and 2. [References to books and to journals containing papers on particular aspects of cometary problems are listed in an appendix.]

THE ORIGIN OF COMETS

RICHTER, N. B. *The nature of comets* (Methuen, 1963).

LYTTLETON, R. A. *The comets and their origin* (Cambridge University Press, 1953), Chs. 3 and 4.

MIDDLEHURST, B. M. *and* KUIPER, G. P. (editors). *The moon, meteorites, and comets: the solar system*, Vol. 4 (Chicago University Press, 1963). [This volume contains large numbers of references to literature on several cometary problems such as the statistics of comets and the production of tails.]

TEKTITES

BAKER, G. Tektites. *Memoirs of the National Museum of Victoria, Melbourne* (1959).

BARNES, V. E. Tektites. *Scientific American* 205 (1961).

NININGER, H. H. *Out of the sky* (Denver University Press, 1952).

O'KEEFE, J. A. (editor). *Tektites* (Chicago University Press, 1963). [Most of the literature of tektites as yet occurs only as reports and papers scattered through numerous scientific journals. Detailed references to a large number of these are given in this volume.]

THE DISCOVERY OF NEPTUNE

ADAMS, J. C. The relevant papers of Adams relating to Neptune are most readily accessible in *The scientific papers of John Couch Adams*, ed. W. G. Adams, Vol. 1 (Cambridge University Press, 1896).

LEVERRIER, U. J. J. The original papers expounding his method are to be found in the *Comptes Rendus* of the Academie des Sciences, Paris, over the years 1839 to 1846. Numerous further references are given in the volume by M. Grosser mentioned below.

BROWN, E. W. On a criterion for the prediction of an unknown planet. *Monthly Notices of the Royal Astronomical Society*, **92**, 80 (1931).

LITTLEWOOD, J. E. A *mathematician's miscellany* (Meuthen, 1953), pp. 117–134.

LYTTLETON, R. A. A short method for the discovery of Neptune. *Monthly Notices of the Royal Astronomical Society*, **118**, 551 (1958).

———, The rediscovery of Neptune. *Vistas in Astronomy*, Vol. 3 (Pergamon Press, 1959).

GROSSER, MORTON. *The discovery of Neptune* (Harvard University Press, 1962). [This volume, by a historian of science, supplies a non-technical account of the whole episode both prior to and subsequent to the discovery, and also an evaluation of the influences of the personalities of the main participants. The book appends an extensive bibliography of primary sources, technical papers, reviews, letters, and subsequent studies relating to the problem.]

Index